SpringerBriefs in Applied Sciences and Technology

Continuum Mechanics

Series editors

Holm Altenbach, Institut für Mechanik, Lehrstuhl für Technische Mechanik, Otto von Guericke University Magdeburg, Magdeburg, Sachsen-Anhalt, Germany
Andreas Öchsner, Griffith School of Engineering, Griffith University, Southport, QLD, Australia

These SpringerBriefs publish concise summaries of cutting-edge research and practical applications on any subject of Continuum Mechanics and Generalized Continua, including the theory of elasticity, heat conduction, thermodynamics, electromagnetic continua, as well as applied mathematics.

SpringerBriefs in Continuum Mechanics are devoted to the publication of fundamentals and applications, presenting concise summaries of cutting-edge research and practical applications across a wide spectrum of fields. Featuring compact volumes of 50 to 125 pages, the series covers a range of content from professional to academic.

More information about this series at http://www.springer.com/series/10528

Marcus Aßmus

Structural Mechanics of Anti-Sandwiches

An Introduction

Springer

Marcus Aßmus
Institute of Mechanics
Otto von Guericke University
Magdeburg, Saxony-Anhalt, Germany

ISSN 2191-530X ISSN 2191-5318 (electronic)
SpringerBriefs in Applied Sciences and Technology
ISSN 2625-1329 ISSN 2625-1337 (electronic)
SpringerBriefs in Continuum Mechanics
ISBN 978-3-030-04353-7 ISBN 978-3-030-04354-4 (eBook)
https://doi.org/10.1007/978-3-030-04354-4

Library of Congress Control Number: 2018962119

This Springer imprint is published by the registered company Springer Nature Switzerland AG
The registered company address is: Gewerbestrasse 11, 6330 Cham, Switzerland

Preface

Cutting-edge engineering applications can also disclose new problems. Especially with structural mechanics, problems arise since computations that follow classical approaches often fail, at least when basic assumptions of underlying theories are violated. Therefore, advanced recipes are needed. This is also the case to what is known as Anti-Sandwich. Due to the slenderness of such structures, they can be assigned to the group of thin-walled structural elements. Anti-Sandwiches comprise multiple layers that allow a classification to the genus of composite structures. However, the discontinuity of mechanical properties in transverse direction is extraordinarily strong. At the same time, the structural thickness of the layers differs widely. Hereby, we leave the framework of classical composite structures. Present elaboration is therefore dedicated to the mechanics of such a species.

This contribution is based on the dissertation of the author. To open the topic to a wider readership, this work was translated and is presented in compact form here. The author thanks Prof. Holm Altenbach for the inclusion in his team and scientific accompaniment during the last three years. The freedom to set up priorities independently was a privilege. Apl. Prof. Konstantin Naumenko is acknowledged for both, confusion and clarification. Obviously not to the same extent.

The author is indebted to Prof. Andreas Öchsner and Prof. Victor A. Eremeyev for beneficial suggestions and encouragement. Additional acknowledgments go to my colleagues Joachim Nordmann, Stefan Bergmann, and Priv.-Doz. Rainer Glüge, which generously answered questions, criticized various parts of this work, and gave valuable comments which improved both, contents and style.

The author is sincerely grateful to Christoph Baumann from Springer-Verlag for assistance and support during the publication process of this book. Throughout the preparation of this work, the author was fellow of the research training group *Micro-Macro-Interactions of Structured Media and Particle Systems* of the German Research Foundation, whose support he very gratefully acknowledges.

Magdeburg, Germany
September 2018

Marcus Aßmus

Contents

Chapter 1
Introduction

1.1 What is an Anti-Sandwich?

Laminates, Sandwiches and Anti-Sandwiches are classically classified as composite structures. Composite structures are multi-layered thin-walled structural elements which exhibit special geometrical features. For this purpose, plane dimensions $L_\alpha \ \forall \alpha \in \{1, 2\}$ and the overall thickness H are used.

$$L_1 \approx L_2 \qquad \wedge \qquad L_\alpha \gg H \tag{1.1}$$

Often, a measure for the slenderness of such structures is introduced. In [8], different significant values are given for the slenderness ratio Ξ. In principle, however, we can give the following statement.

$$\Xi = \frac{H}{L_{\min}} \ll 1 \qquad L_{\min} = \min\{L_1, L_2\} \tag{1.2}$$

Since we will deal with symmetric three-layered composites where the skin layers have identical geometrical and physical properties, the following relationship holds for the thickness.

$$H = 2h^s + h^c \tag{1.3}$$

Herein, $h^K \ \forall K \in \{s, c\}$ are the thicknesses of core (c) and skin layers (s). The decision whether it is a Sandwich or an Anti-Sandwich can now be made via the correlations of the individual layers.

$$h^s \ll h^c \qquad \text{Sandwich} \tag{1.4}$$

$$h^s \gg h^c \qquad \text{Anti-Sandwich} \tag{1.5}$$

M. Aßmus, *Structural Mechanics of Anti-Sandwiches*,
SpringerBriefs in Continuum Mechanics,
https://doi.org/10.1007/978-3-030-04354-4_1

For laminates $h^s \approx h^c$ holds. In context of this interpretation we can define an Anti-Sandwich as geometrically contrary compared to a classical Sandwich structure. However, strongly diverging material properties also cause problems when they reach a certain threshold. Therefore, the shear modulus G is often used for evaluation. In the case of isotropy, following relation holds true for Sandwiches and Anti-Sandwiches.

$$G^c \ll G^s \qquad G^K = \frac{Y^K}{2(1+\nu^K)} \quad \forall K \in \{s, c\} \tag{1.6}$$

Herein, Y is Young's modulus and ν is Poisson's ratio. For laminates $G^c \approx G^c$ holds. In relation with geometrical features, the shear modulus is also used to describe the shear rigidity D_S.

$$D_S^K = \kappa^K G^K h^K \qquad \forall K \in \{s, c\} \tag{1.7}$$

Herein, κ is the so-called shear correction factor. To conclude, following relation holds for Sandwiches and Anti-Sandwiches, even while considering relations (1.4) and (1.5).

$$D_S^c \ll D_S^s \tag{1.8}$$

An illustration of the circumstances described above is given in Fig. 1.1. Here, a direct juxtaposition of Sandwich and Anti-Sandwich elucidate the definition introduced for the latter one.

One of the first mentions of this special species can be found in a paper of Altenbach et al. [5] whereby also earlier mentions can be found, which do not necessarily treat identical problem.

1.2 Why a Special Approach?

There are various structural mechanics models which can be found in literature to analyze composite structures. The most popular approach is based on the first-order shear deformation theory [3]. It is barely impossible to summarize the amount of available theories and calculation methods in context of present treatise. In general one can name layerwise and zig-zag theories as extended approaches. Carrera [10, 11] attempted to recapitulate these areas. However, these reviews were compiled more than 15 years ago. In the meantime, due to the wide variety of further developments, they are no longer up to date. The number of publications in this field appears to be almost unmanageable. Recent attempts to give overviews fail not least because, in addition to the English language, basic knowledge of the French, Russian, Italian and German language is required in which significant innovations are published. According to the opinion of the author, such a summary should therefore be of compendial nature.

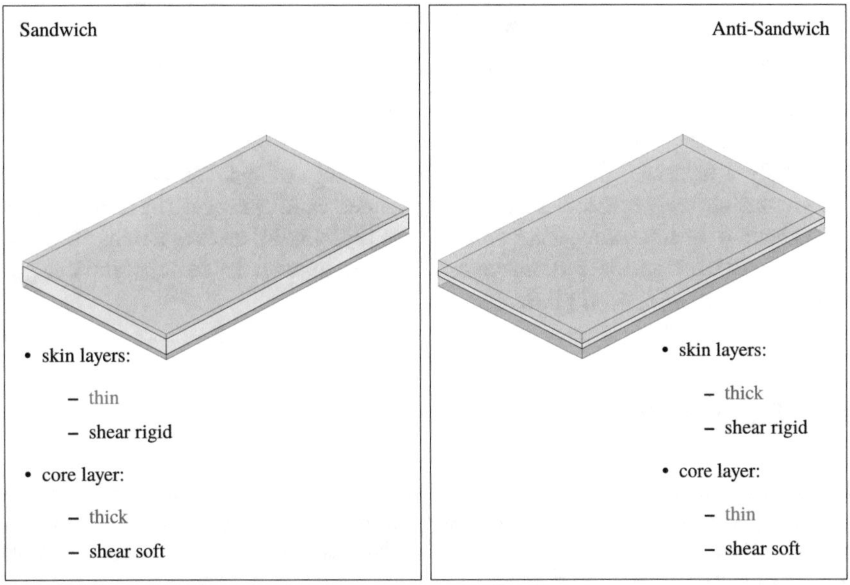

Fig. 1.1 Sandwich and Anti-Sandwich: Differences and similarities

However, the basic assumptions of the first-order shear deformation theory also incorporates the presumption that material properties and structural thicknesses of the individual layers are in a similar order of magnitude. In contrast, Anti-Sandwiches show strongly diverging material properties and layer thicknesses. For this reason we affiliate Anti-Sandwiches to the more generalized composite structures.

In [36] different measures are introduced to substantiate the necessity of an alternative approach. On the one hand, the shear modulus ratio *GR* was introduced to confirm the extent of validity of first-order shear deformation or sandwich theories [39] based on the works of Reissner [43, 44] and Mindlin [34]. This range is given as follows.

$$GR = \frac{G^c}{G^s} = 10^{-2} \ldots 10^{-1} \tag{1.9}$$

Furthermore, a dimensionless parameter was introduced which incorporates geometrical influences also [36]. In the context of symmetric Anti-Sandwiches this geometrically normalized shear rigidity parameter is given as follows.

$$NGRP = \sqrt{\frac{2\left[1-(\nu^s)^2\right]\kappa^c G^c L_{\min}^2}{Y^s h^s h^c}\left[1+3\left(1+\frac{h^c}{h^s}\right)^2\right]} \tag{1.10}$$

In [17] we can find various validity ranges for applicable theories.

$$NGRP = 6 \cdot 10^1 \ldots 10^2 \quad \text{Kirchhoff theory} \tag{1.11}$$
$$NGRP = 2 \cdot 10^1 \ldots 10^2 \quad \text{first-order shear deformation theory} \tag{1.12}$$
$$NGRP = 4 \cdot 10^{-2} \ldots 10^2 \quad \text{layerwise theory} \tag{1.13}$$

Obviously, a layerwise theory shows the widest range of application while classical theories are strongly restrictive within the bounded validity range. The term classical refers here to the first-order shear deformation [34, 43, 44] and the Kirchhoff theory [28]. Due to these special circumstances, classical approaches to structural analysis as summarized in [2, 4, 41] fail.

1.3 Frame of Reference

From standpoint of classical engineering, the fundamentals of (multi-layered) thin-walled structures are laid in the papers of Kirchhoff [28], Reissner [44], and Mindlin [34]. While these three are well known and most cited, Föppl [20], von Kármán [52], Aron [6], Love [33], and Hencky [25] should also be mentioned.

In the context of these shell and plate theories Timoshenko and Woinowsky-Krieger [48], Girkmann [21], Başar and Krätzig [8], Altenbach, Altenbach and Naumenko [3], Reddy [42], and Sab and Lebée [46] provide summaries. Of course, the comprehensive source of Naghdi [35] has not lost importance today.

The concept when operating with shells and plates is to reduce all considerations to a surface within (or beyond) the body observed. All further procedures are restricted to this surface. The derivation of such dimensional reduced continua is still controversial, cf. discussion in [29]. In principle we can divide between four different strategies. Three of them are based on a three dimensional Cauchy continuum (also known as Boltzmann continuum):

- the asymptotic approach [13, 22, 26]
- the hypothetic approach [25, 28, 31, 34, 44]
- the (pseudo-)consistent approach [12, 27, 40].

Following Neff [37], none of these three approaches is consistent. This leads us to the fourth approach which is based on a surface continuum. This is called

- the direct approach [14, 15, 23, 53].

This direct access is corroborated by Truesdell [49]. However, divergent partitions of above fourfold division can be found in [1] or [38], for example.

Following Truesdell and Toupin [50], surface continua in the sense of the Cosserat brothers belong to the so-called oriented bodies. These micro-polar continua are usually associated with the name Cosserat. Even if the Cosserat brothers were the godfathers of the notion to introduce independent rotational degrees of freedom, cf. [15], this idea was previously published by others. As stated in [49], Leonhard Euler already recognized the conservation of momentum and the conservation of angular

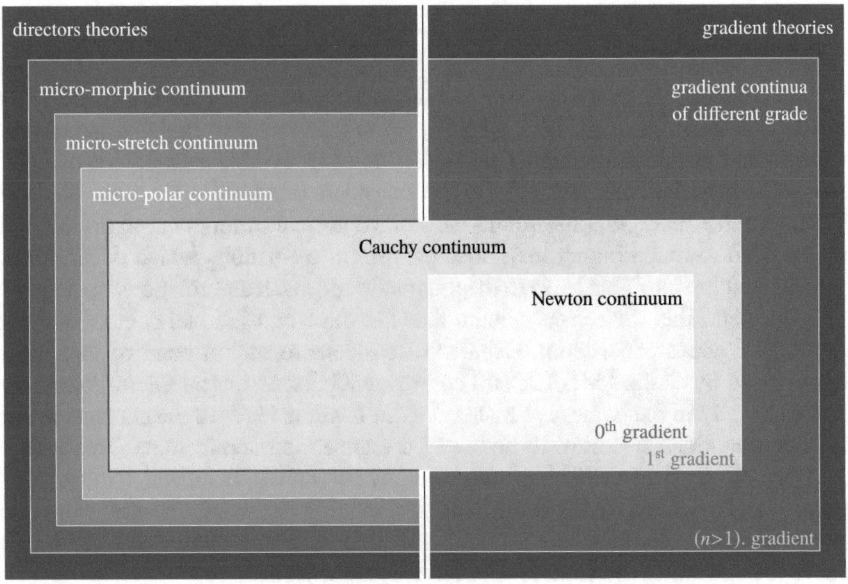

Fig. 1.2 Context of generalized continuum theories based on the two principle movements

momentum as two independent, fundamental principles. Even Jacob Bernoulli postulated the angular momentum independent of momentum balance. Voigt had such ideas in crystal elasticity [51]. Also Duhem [16] noted that certain phenomena are not reproducible with the (now classical) Cauchy continuum [12]. However, long unnoticed by the scientific community, the Cosserat's work experienced a revival in the late fifties of the last century initiated by Ericksen and Truesdell [18], see [24] or [47].

However, following Eringen [19] this micro-polar theory belongs to the so-called directors theories. In Fig. 1.2 these theories are visualized at the left-hand side. Subsets are micro-morphic, micro-stretch, and micro-polar continuum. This class of theories is assigned to the generalized continuous. On the opponent side of this figure are the so-called gradient theories [9]. Both movements are apparently the most popular nowadays, where literature can show further movements too. In the framework of these generalized theories, the Cauchy continuum is just a special case. Deviating and expanded overviews for the classification of generalized continua are given in [19, 30, 45].

Finally we want to emphasize the red line between Cauchy continuum and micro-polar continuum. Engineers usually cross that red line bidirectional [32] when deducing a two dimensional continuum and restoring quantities of a three dimensional body manifold. However, since such a procedure causes inconsistencies in stress and strain recovery, we reduce our concern to the mechanics of surfaces solely.

1.4 Concept and Structure

In the context of the present work the principle of Truesdell [49] is pursued and a surface continuum is introduced, whereby the independence of impulse and angular momentum balance is assumed a priori. It follows from this that rotations independent of the translations must exist. If one considers rotations at a surface element, it is noticeable that rotations around the surface normal (orthogonal to the surface) are exposed to much bigger resistance compared to rotations which occur during bending and torsion. This leads to the pragmatic approach that rotations in the plane are neglected. This surface continuum thus has three translational degrees and two rotational degrees of freedom. Detailed discussions to such a point of view were summarized by Zhilin [54]. The still controversial discussion on micro-polar theories, especially in the context of material identification, leads to the conclusion that tangible quantities still have to be used for engineering applications. The basis is therefore mostly the three dimensional Cauchy continuum, from which this consideration can be derived. In the sense of contextualizing Fig. 1.2 the present work thus moves on the border between Cauchy and micro-polar continuum. The red line in this figure illustrates the work area of the present considerations.

Even if these abstract constructs of generalized continuum theories increasingly find their way into scientific literature, such a consideration is not always advantageous in practice. In the context of a two dimensional continuum, a critical definition according to Libai and Simmonds [32] can be given as follows.

There is no physical, two dimensional surface. It is a pure theoretical construct for practicable mathematical handling, derived from our three dimensional world.

In this sense, we consider the surface as primitive concept. In present context, the notion of an elastic surface is just an illustrative means for the stretching, shearing, bending, and twisting of a single layer, as we will see later.

Classically in engineering sciences first a transformation of a three dimensional Cauchy continuum into a surface continuum of the genus presented here is made. In the context of a consistent derivation problems arise [7]. Therefore the basic equations are mostly constructed rigorously in the sense of an engineering approach. In the present work we labour without such derivations. But also the restriction is connected with it. These are related with a procedure well known as stress and strain recovery.

While starting with a surface continuum we limit ourselves to a completely linear theory. This means linear material behavior with small deformations (translations and rotations). This approach is practical and formally simpler than formulating a geometrically nonlinear theory right from the start. Ultimately, this approach should only allow a correlation to the quantities manifested in engineering sciences, which, however, is no less problematic than a direct approach.

The explanations to the direct approach are mainly based on the textbook of Zhilin [54]. The basic assumption of this direct formulation is that the continuum can be represented by a deformable surface in three dimensional space. The advantage

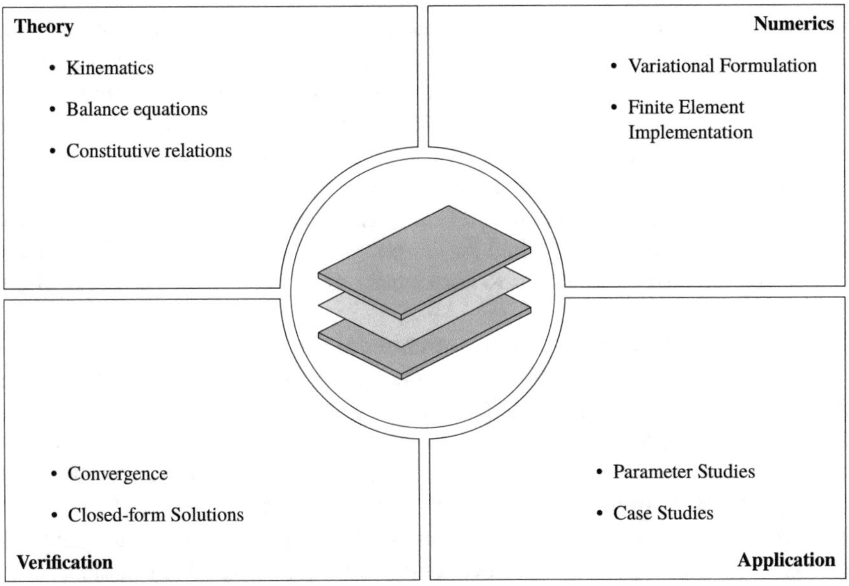

Fig. 1.3 Procedure for present considerations and structure of the work

of this theory is that its derivation does not depend on hypotheses. It is formulated mathematically and physically exact and as exact as the three dimensional continuum mechanics.

In summary, the considerations are limited to the following aspects:

- flatness of the structure (initially uncurved)
- uniform layer thickness of the individual layers
- geometric linearity (small displacements and small rotations)
- physical linearity (linear elastic material behavior)
- independence of translations and rotations.

Couplings with thermodynamics, electrodynamics and other non-mechanical influences are neglected in the present treatise. Likewise, non-local effects or dependencies of higher gradients are not considered. Even though we limit ourselves to apolar simple materials where a deterministic relation between local deformations and symmetrical stresses is assumed, this condition is loosened and independent rotations are introduced. For reasons of simplicity, orientation dependencies should be excluded as well as physical nonlinearities.

The descriptions are divided as follows, whereby the central aspects are visualized in Fig. 1.3.

In Chap. 2 the basic equations of planar surface continua are introduced canonically and thus the attempt is made to present the theory of planar surface continua axiomatically. In addition the claim is followed to introduce a geometrically exact theory. Analogous to the procedure with classical continua, kinematic and kinetic measures are defined and constitutive relations are discussed based on the description of the position and the introduction of degrees of freedom. The chapter concludes with a clear overall formulation of the model for a planar surface continuum.

Within the scope of the further explanations the ideas in the sense of Naumenko and Eremeyev [36] are extended to multilayered structures in Chap. 3. Hereby we introduce a more generalized approach. The assumptions and consequences of the introduced constraints are discussed for the coupling of multiple surface continua which are required for this purpose. Furthermore, significant global kinetic and kinematic quantities are introduced and balanced over the overall structure.

In general, the initial boundary value problem of elastostatics of multi-layer planar surface continua cannot be solved analytically. Chapter 4 therefore introduces a variation formulation for the generation of the weak form of equilibrium, based on the principle of virtual work. In the following Chap. 5 follows the transition to a discrete solution of the mechanical problem by a numerical solution approach. The implementation of the underlying equations into a finite element program system is presented. Here the vector-matrix notation proves to be efficient with regard to a compact notation, especially in context to the programmatic implementation.

Within Chap. 6 follows the verification of the numerical solution generated for a representative model structure. First, convergence studies are carried out, the performance of the approach shown is discussed, and then comparative solutions are used for the replication. Thereby closed-form solutions for limiting cases of the mechanical behaviour are used.

In Chap. 7 application-related investigations on an Anti-Sandwich are presented. First, parameter studies for the variation of geometric and material quantities of the structure are carried out and directions for optimal parameters are shown. Subsequently, an exemplary structure is used to expose it to more realistic loads. Two characteristic weathering scenarios are used for this purpose. The results of this structural analysis are discussed in detail.

The work concludes with a summary of the central aspects, giving impulses for possible further topics.

1.5 A Note on Notational Conventions

In present work a direct notation is preferred. This is useful for avoiding confusion and laborious representations of the governing equations as it is often the case by using indicial notation. However, the following simple concept for the notation is applied where the letter 'a' is used in different style exemplary.

a	scalars,	tensor of 0th order	(— , inclined, normal font weight)
$\boldsymbol{a}$	monads,	tensor of 1st order	(minuscule, inclined, bold)
$\boldsymbol{A}$	dyads,	tensor of 2nd order	(majuscule, inclined, bold)
$\boldsymbol{\mathcal{A}}$	tetrads,	tensor of 4th order	(majuscule, inclined, bold, calligraphic)

a	vector	(minuscule, upright, bold, sans serif)
A	matrix	(majuscule, upright, bold, sans serif)
$\mathbb{A}$	sets, groups, and spaces	(blackboard bold)
$\mathfrak{A}$	material body manifolds	(gothic print – black letter)
$\mathcal{A}$	functions	(majuscule, inclined, calligraphic)

If we indulge in indicial notation, latin indices run through the values 1, 2, and 3, while greek indices run through the values 1 and 2 only. Deviations thereof are indicated in the text.

The reader should be familiar with the basics of calculus and analysis. However, the main part of this work is written in modern tensor notation. A brief introduction to mathematical operations of tensors is given in Appendix A. Hopefully the reader will endure this excurse.

References

1. Altenbach H (1998) Theories for laminated and sandwich plates. Mech Compos Mater 34(3):243–252. https://doi.org/10.1007/BF02256043
2. Altenbach H, Altenbach J, Rikards R (1996) Einführung in die Mechanik der Laminat- und Sandwichtragwerke - Modellierung und Berechnung von Balken und Platten aus Verbundwerkstoffen. Deutscher Verlag für Grundstoffindustrie, Stuttgart
3. Altenbach H, Altenbach J, Naumenko K (1998) Ebene Flächentragwerke: Grundlagen der Modellierung und Berechnung von Scheiben und Platten. Springer, Berlin. https://doi.org/10.1007/978-3-642-58721-4
4. Altenbach H, Altenbach J, Kissing W (2004) Mechanics of composite structural elements. Springer, Berlin. https://doi.org/10.1007/978-3-662-08589-9
5. Altenbach H, Eremeyev VA, Naumenko K (2015) On the use of the first order shear deformation plate theory for the analysis of three-layer plates with thin soft core layer. Zeitschrift für Angewandte Mathematik und Mechanik 95(10):1004–1011. https://doi.org/10.1002/zamm.201500069
6. Aron H (1874) Das Gleichgewicht und die Bewegung einer unendlich dünnen, beliebig gekrümmten elastischen Schale. Journal für die reine und angewandte Mathematik 78:136–174. https://doi.org/10.1515/crll.1874.78.136
7. Aßmus M (2018) Global structural analysis at photovoltaic modules: theory, numerics, application (in German). Dissertation, Otto von Guericke University Magdeburg
8. Başar Y, Krätzig WB (1985) Mechanik der Flächentragwerke: Theorie Berechnungsmethoden, Anwendungsbeispiele. Springer, Wiesbaden. https://doi.org/10.1007/978-3-322-93983-8
9. Bertram A (2016) Compendium on gradient materials. Otto-von-Guericke Universität, Magdeburg. http://www.ifme.ovgu.de/ifme_media/FL/Publikationen/Compendium+on+Gradient+Materials+Okt+2016.pdf
10. Carrera E (2002) Theories and finite elements for multilayered, anisotropic, composite plates and shells. Arch Comput Methods Eng 9(2):87–140. https://doi.org/10.1007/BF02736649
11. Carrera E (2003) Theories and finite elements for multilayered plates and shells: a unified compact formulation with numerical assessment and benchmarking. Arch Comput Methods Eng 10(3):215–296. https://doi.org/10.1007/BF02736224
12. Cauchy AL (1828) Sur l'équilibre et le movement intérieur des corps considérés comme des masses continues. Ex de Math 4:293–319. http://catalogue.bnf.fr/ark:/12148/cb302073189
13. Ciarlet PG (1990) Plates and junctions in elastic multi-structures: an asymptotic analysis, vol 14. Research in applied mathematics. Masson, Paris

14. Cohen H, DeSilva CN (1966) Nonlinear theory of elastic directed surfaces. J Math Phys 7(6):960–966. https://doi.org/10.1063/1.1705009
15. Cosserat E, Cosserat F (1909) Théorie des corps déformables. A. Hermann et fils, Paris. http://jhir.library.jhu.edu/handle/1774.2/34209
16. Duhem P (1893) Le potentiel thermodynamique et la pression hydrostatique. Annales scientifiques de l'École Normale Supérieure 10:183–230. https://doi.org/10.24033/asens.389
17. Eisenträger J, Naumenko K, Altenbach H, Meenen J (2015) A user-defined finite element for laminated glass panels and photovoltaic modules based on a layer-wise theory. Compos Struct 133:265–277. https://doi.org/10.1016/j.compstruct.2015.07.049
18. Ericksen JL, Truesdell C (1957) Exact theory of stress and strain in rods and shells. Arch Rat Mech Anal 1(1):295–323. https://doi.org/10.1007/BF00298012
19. Eringen AC (1999) Microcontinuum field theories: I. Foundations and solids. Springer, New York. https://doi.org/10.1007/978-1-4612-0555-5
20. Föppl A (1907) Vorlesungen über technische Mechanik. B.G Teubner, Leipzig
21. Girkmann K (1986) Flächentragwerke – Einführung in die Elastostatik der Scheiben, Platten, Schalen und Faltwerke, 6th edn. Springer, New York (First Edition 1946)
22. Goldenweizer A (1962) Formulation of approximative theory of shells with the help of the asymptotic integration of the equations of the theory of elasticity (in Russian). Prikl Mat i Mekh 26(4):668–686
23. Green AE, Naghdi PM, Wainwright WL (1965) A general theory of a cosserat surface. Arch Rat Mech Anal 20(4):287–308. https://doi.org/10.1007/BF00253138
24. Günther W (1958) Zur Statik und Kinematik des Cosseratschen Kontinuums. Abhandlungen der Braunschweigischen Wissenschaftlichen Gesellschaft 10:195–213. https://publikationsserver.tu-braunschweig.de/receive/dbbs_mods_00046248
25. Hencky H (1947) Über die Berücksichtigung der Schubverzerrung in ebenen Platten. Ingenieur-Archiv 16(1):72–76. https://doi.org/10.1007/BF00534518
26. Kaplunov JD, Kossovich LY, Nolde E (1998) Dynamics of thin walled elastic bodies. Academic Press, Cambridge
27. Kienzler R, Schneider P (2012) Consistent theories of isotropic and anisotropic plates. J Theor Appl Mech 50(3):755–768
28. Kirchhoff GR (1850) Über das Gleichgewicht und die Bewegung einer elastischen Scheibe. Journal für die reine und angewandte Mathematik 40:51–88. https://doi.org/10.1515/crll.1850.40.51
29. Koiter W (1969) Foundations and basic equations of shell theory: a survey of recent progress. Theory of thin shells. IUTAM symposium copenhagen 1967. Springer, Heidelberg, pp 93–105. http://www.springer.com/gp/book/9783642884788
30. Kröner E (1968) Interrelations between various branches of continuum mechanics. Springer, Berlin, pp 330–340. https://doi.org/10.1007/978-3-662-30257-6_40
31. Levinson M (1980) An accurate, simple theory of the statics and dynamics of elastic plates. Mech Res Commun 7(6):343–350. https://doi.org/10.1016/0093-6413(80)90049-X
32. Libai A, Simmonds JG (1983) Nonlinear elastic shell theory. Adv Appl Mech 23:271–371. https://doi.org/10.1016/S0065-2156(08)70245-X
33. Love AEH (1888) The small free vibrations and deformation of a thin elastic shell. Philos Trans R Soc Lond A: Math, Phys Eng Sci 179:491–546. https://doi.org/10.1098/rsta.1888.0016
34. Mindlin RD (1951) Influence of rotatory inertia and shear on flexural motions of isotropic, elastic plates. J Appl Mech 18:31–38
35. Naghdi PM (1972) The theory of shells and plates. In: Flügge W (ed) Encyclopedia of physics - linear theories of elasticity and thermoelasticity, vol VI, a/2 (ed. C. Truesdell). Springer, Berlin, pp 425–640. https://doi.org/10.1007/978-3-662-39776-3_5
36. Naumenko K, Eremeyev VA (2014) A layer-wise theory for laminated glass and photovoltaic panels. Compos Struct 112:283–291. https://doi.org/10.1016/j.compstruct.2014.02.009
37. Neff P, Hong KI, Jeong J (2010) The Reissner-Mindlin plate is the Γ-limit of Cosserat elasticity. Math Model Methods Appl Sci 20(9):1553–1590. https://doi.org/10.1142/S0218202510004763

38. Noor AK, Burton WS (1989) Assessment of shear deformation theories for multilayered composite plates. Appl Mech Rev 42(1):1–13. https://doi.org/10.1115/1.3152418
39. Planterna F (1966) Sandwich construction: the bending and buckling of sandwich beams, plates, and shells. Wiley, New York
40. Preußer G (1982) Eine Erweiterung der Kirchhoffschen Plattentheorie. Dissertation, Technische Hochschule Darmstadt
41. Reddy JN (2004) Mechanics of laminated composite plates and shells: theory and analysis, 2nd edn. Taylor & Francis, Boca Raton
42. Reddy JN (2006) Theory and analysis of elastic plates and shells, 2nd edn. Taylor & Francis, Boca Raton
43. Reissner E (1944) On the theory of bending of elastic plates. J Math Phys 23(1–4):184–191. https://doi.org/10.1002/sapm1944231184
44. Reissner E (1945) The effect of transverse shear deformation on the bending of elastic plates. J Appl Mech 12:69–77
45. Saanouni K (2012) Damage mechanics in metal forming: advanced modeling and numerical simulation. ISTE Ltd and Wiley, London. https://doi.org/10.1002/9781118562192
46. Sab K, Lebeé A (2015) Homogenization of heterogeneous thin and thick plates. Mechanical engineering and solid mechanics series. ISTE Ltd and Wiley, London. https://doi.org/10.1002/9781119005247
47. Schäfer M (1962) Versuch einer Elastizitätstheorie des zweidimensionalen ebenen Cosserat-Kontinuums. Akademie, Berlin, pp 277–292
48. Timoshenko S, Woinowsky-Krieger S (1987) Theory of plates and shells, 2nd edn. McGraw-Hill, New York (First Edition 1959)
49. Truesdell C (1964) Die Entwicklung des Drallsatzes. Zeitschrift für Angewandte Mathematik und Mechanik 44(4–5):149–158. https://doi.org/10.1002/zamm.19640440402
50. Truesdell C, Toupin RA (1960) The classical field theories. In: Flügge S (ed) Encyclopedia of physics - principles of classical mechanics and field theory, vol 2/3/1. Springer, Berlin, pp 226–858. https://doi.org/10.1007/978-3-642-45943-6_2
51. Voigt W (1887) Theoretische Studien über die Elasticitätsverhältnisse der Krystalle. Abhandlungen der Königlichen Gesellschaft der Wissenschaften in Göttingen 34:3–52. http://eudml.org/doc/135896
52. von Kármán T (1910) Festigkeitsprobleme im Maschinenbau, vol IV. Encyklopädie der mathematischen Wissenschaften, pp 311–384
53. Zhilin PA (1976) Mechanics of deformable directed surfaces. Int J Solids Struct 12(9):635–648. https://doi.org/10.1016/0020-7683(76)90010-X
54. Zhilin PA (2006) Applied mechanics - foundations of shells theory (in Russian). Publisher of the Polytechnic University, St. Petersburg. http://mp.ipme.ru/Zhilin/Zhilin_New/pdf/Zhilin_Shell_Book.pdf

Chapter 2
Theory of Planar Surface Continua

2.1 Planar Continua

Planar continua analogous to the direct approach will be introduced at this point, i.e. we will operate on a deformable surface instead of a voluminous body a priori. The description is following the Zhilinean path [11] though using a more common and unique notation. However, the main hypotheses of the classical mechanics of continuous media hold true. The material surface $\mathfrak{S}$ is a coherent and compact set of material space points $\mathfrak{M}$. The boundary of this point set is indicated by the domain boundary $\partial\mathfrak{S}$. First, let us start with a homogeneous body, i.e. all material points have the same characteristics. Another limitation is in the isotropy assumption, i.e. all directions are equal. Each material point has three translational and two rotational degrees of freedom as kinematic variables, with the rotations introduced as independent degrees of freedom. The continuum hypothesis applies, maintaining the continuity of material points during deformation. The relationship to the three dimensional body $\mathfrak{B}$ in whose volume V the surface is embedded can be represented by the following expression, where the assumption $h = \text{const.}$ holds for the structural thickness.

$$V = \left\{(X_1, X_2, X_3) \in \mathfrak{B} \subset \mathbb{E}^3 : (X_1, X_2) \in \mathfrak{S} \subset \mathbb{E}^2, X_3 \in [-{}^{h}\!/\!_{2}, +{}^{h}\!/\!_{2}]\right\} \tag{2.1}$$

The compact manifold $\mathfrak{S}$ is a subset of the two dimensional subspace $\mathbb{E}^2$ embedded into the Euclidean space $\mathbb{E}^3$. It should be pointed out that all material points of present surface continuum are coplanar, i.e. lying on a plane. The boundary of the surface continuum $\partial\mathfrak{S}$ can be subdivided in an boundary at $\pm\boldsymbol{n}$ and in boundaries at the plane borders $\pm\boldsymbol{e}_\alpha$, here with the boundary normal $\boldsymbol{\nu}$.

$$\partial\mathfrak{S} = \partial\mathfrak{S}^+ \cap \partial\mathfrak{S}^- \cap \partial\mathfrak{S}^\nu \tag{2.2}$$

M. Aßmus, *Structural Mechanics of Anti-Sandwiches*,
SpringerBriefs in Continuum Mechanics,
https://doi.org/10.1007/978-3-030-04354-4_2

Here $\boldsymbol{n}$ is the normal to the surface basis and $\boldsymbol{\nu}$ is the normal to the surface contours. Subsequently, basic concepts for describing and analyzing the position and the movement of the material surface are introduced.

2.2 Kinematics

First, the surface continuum now defined should be embedded in a space. The space of physical intuition is the Euclidean space, because it has measures for angles and lengths. In addition to these spatial measures, a temporal description is available for the sequence of configurations or placements. Thus, a time-dependent embedding in the three dimensional Euclidean space $\mathbb{E}^3$ can be used to describe movements. For every point in time t of such an embedding, the term placement is used.

2.2.1 Reference and Current Placement

The embedding of the set of material points $\mathfrak{M}$ in the Euclidean space $\mathbb{E}^3$ is done by choosing a time-invariant, space-constrained, Cartesian frame of reference $\{\boldsymbol{i}_1, \boldsymbol{i}_2, \boldsymbol{i}_3\}$. At any time, this process can be described as follows.

$$\Lambda_t := \mathfrak{S} \rightarrow \mathbb{E}^3 \tag{2.3}$$

At current time the surface embedded is defined by the following expression.

$$\mathfrak{S}_t := \Lambda(\mathfrak{S}, t) \subset \mathbb{E}^3 \tag{2.4}$$

The embedding of an arbitrary reference placement $\mathfrak{S}_0$ at a freely chosen time t_0 is then introduced as follows.

$$\Lambda_0 := \mathfrak{S}_0 \rightarrow \mathbb{E}^3 \tag{2.5}$$

The space occupied by the surface continuum can then be described as follows.

$$\mathfrak{S}_0 := \Lambda(\mathfrak{S}, t_0) \subset \mathbb{E}^3 \tag{2.6}$$

Furthermore, each placement is assigned with a local, orthogonal, point-fixed base $\{\boldsymbol{e}_1, \boldsymbol{e}_2, \boldsymbol{n}\}$. As part of the restriction to small deformations, the coordinate systems $\{\boldsymbol{e}_\alpha, \boldsymbol{n}\}$ and $\{\boldsymbol{e}'_\alpha, \boldsymbol{n}'\}$ coincide. The normal vector is defined as follows.

$$\boldsymbol{n} = \boldsymbol{e}_1 \times \boldsymbol{e}_2 \tag{2.7}$$

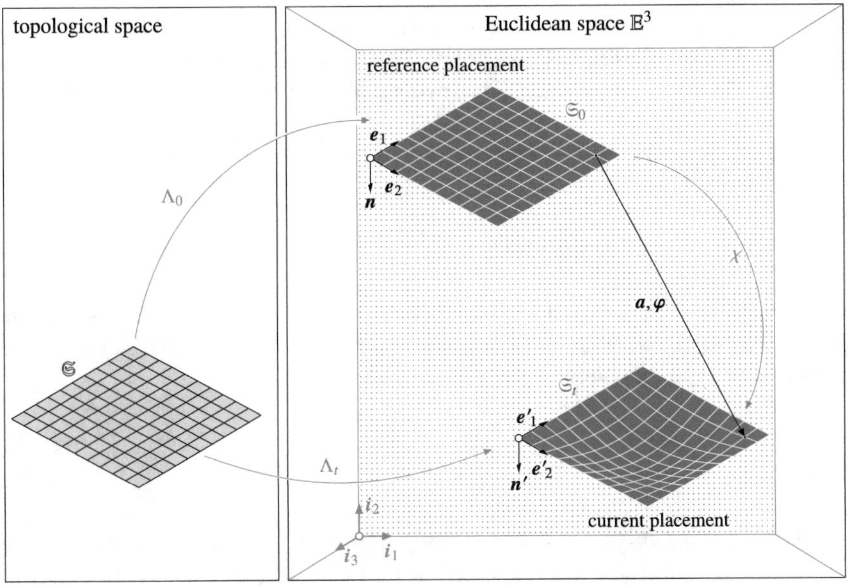

Fig. 2.1 Embedding and kinematics of surface continuum in the Euclidean space

The boundary normal can be determined as follows.

$$\boldsymbol{\nu} = \pm \boldsymbol{e}_2 \times \boldsymbol{n} \qquad \vee \qquad \boldsymbol{\nu} = \pm \boldsymbol{n} \times \boldsymbol{e}_1 \tag{2.8}$$

Following correlation applies for the normals $\boldsymbol{n}$ and $\boldsymbol{\nu}$.

$$\boldsymbol{\nu} \cdot \boldsymbol{n} = 0 \tag{2.9}$$

The visual display of present descriptions is given in Fig. 2.1.

2.2.2 *Degrees of Freedom and Deformation Measures*

The position of a material point of the considered surface in the space $\mathbb{E}^3$ can be described by the position vector as $\boldsymbol{x} = X_i \boldsymbol{i}_i$. By a favorable choice of the reference point, the coordinate systems $\boldsymbol{i}_i = \{\boldsymbol{e}_\alpha, \boldsymbol{n}\}$ can coincide, which simplifies the following explanations. Restricting to the description of the points of the planar surface, the position vector is given by $\boldsymbol{r} = X_\alpha \boldsymbol{e}_\alpha$. The position of the material point in the context of the original volume is determined as follows.

$$\boldsymbol{x} = \boldsymbol{r} + X_3 \boldsymbol{n} \qquad \boldsymbol{r} = X_\alpha \boldsymbol{e}_\alpha \qquad \forall\, \alpha \in \{1, 2\} \tag{2.10}$$

The vector $\boldsymbol{r}$ stands for the position vector of the plane, which is determined by the projection of the three dimensional position vector $\boldsymbol{x} \cdot \boldsymbol{P}$. Herein $\boldsymbol{P} = \boldsymbol{e}_\alpha \otimes \boldsymbol{e}_\alpha$ is the first metric tensor of the surface. Thus every material point is clearly identifiable at any time.

The motion χ of the body in space is given with respect to the introduced reference placement.

$$\chi := \Lambda_t(\Lambda_0^{-1}) : \mathbb{E}^3 \supset \mathfrak{S}_0 \to \mathfrak{S}_t \subset \mathbb{E}^3 \tag{2.11}$$

This motion is here described by the translation vector of the surface $\boldsymbol{a}$ and the physical rotation vector $\boldsymbol{\psi}$. Each material point of the planar continuum is thus endowed with the following degrees of freedom.

$$\boldsymbol{a} = v_1 \boldsymbol{e}_1 + v_2 \boldsymbol{e}_2 + w \boldsymbol{n} \tag{2.12}$$

$$\boldsymbol{\psi} = -\varphi_2 \boldsymbol{e}_1 + \varphi_1 \boldsymbol{e}_2 \tag{2.13}$$

The third rotational degree of freedom – rotation in the plane (φ_3) around the normal $\boldsymbol{n}$ – is neglected. Intuitively, this is justified by the fact that the resistance of the surface is much larger here compared to the resistance to bending or torsion. Based on the Cosserat rotations $\tilde{\boldsymbol{\varphi}} = \varphi_i \boldsymbol{e}_i = \varphi_\alpha \boldsymbol{e}_\alpha + \varphi_3 \boldsymbol{n}$, this two dimensional vector can be are derived as follows.

$$\boldsymbol{\psi} = -\tilde{\boldsymbol{\varphi}} \times \boldsymbol{n} = -\varphi_2 \boldsymbol{e}_1 + \varphi_1 \boldsymbol{e}_2 \tag{2.14}$$

For the rational description of the following equations, the pseudo-rotation vector $\boldsymbol{\varphi}$ is introduced following Pal'mov [7], which is constructed as follows in relation to the physical rotation vector $\boldsymbol{\psi}$. This mathematical rotation vector can be generated by projection of Cosserat rotations.

$$\boldsymbol{\varphi} = \varphi_1 \boldsymbol{e}_1 + \varphi_2 \boldsymbol{e}_2 \qquad \boldsymbol{\varphi} = \tilde{\boldsymbol{\varphi}} \cdot \boldsymbol{P} \tag{2.15}$$

The relationship between both vectors is given as follows.

$$\boldsymbol{\varphi} = \boldsymbol{\psi} \times \boldsymbol{n} \qquad \Leftrightarrow \qquad \boldsymbol{\psi} = -\boldsymbol{\varphi} \times \boldsymbol{n} = \boldsymbol{n} \times \boldsymbol{\varphi} \tag{2.16}$$

The determination of the vector of the translational degrees of freedom of the surface continuum is based on the two plane displacements v_α and the deflection w.

$$\boldsymbol{a} = \boldsymbol{v} + w\boldsymbol{n} \qquad \boldsymbol{v} = v_\alpha \boldsymbol{e}_\alpha \qquad \forall\, \alpha \in \{1, 2\} \tag{2.17}$$

The position of a material point in the current placement is thus a function of translations and rotations.

$$\boldsymbol{r} = \boldsymbol{r}_0 + \mathcal{F}(\boldsymbol{a}, \boldsymbol{\varphi}) \tag{2.18}$$

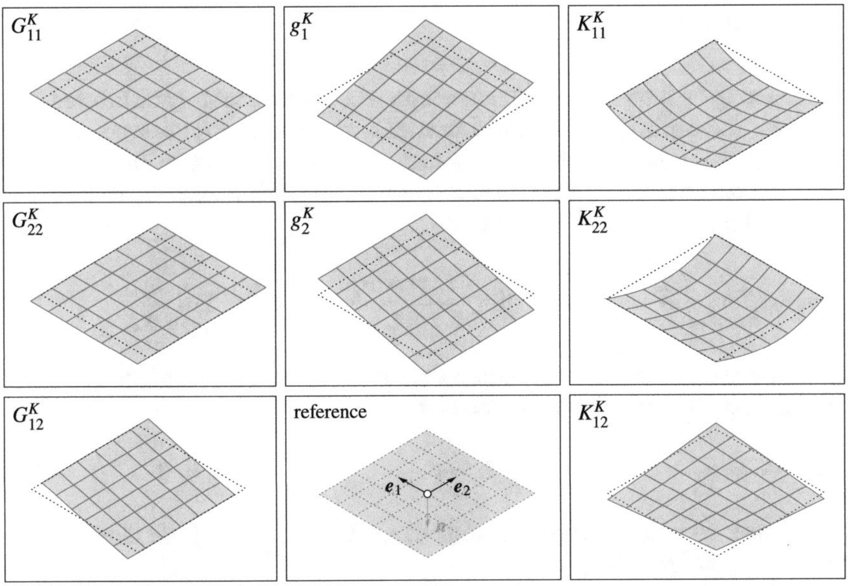

Fig. 2.2 Visualization of representative deformations at an infinitesimal surface element

Similar to the approach in classical continuum mechanics, deformation measures can be introduced using the gradients of the plane displacements $\nabla \boldsymbol{v}$, the deflections ∇w and the rotations $\nabla \boldsymbol{\varphi}$. Terms of higher order should be neglected here. Since rigid body motions should have no influence on kinetic quantities, only the respective symmetrical component is considered. The following deformation measures can thus be defined.

$$\boldsymbol{G} = \nabla^{\text{sym}} \boldsymbol{v} \quad = G_{\alpha\beta}\ \boldsymbol{e}_\alpha \otimes \boldsymbol{e}_\beta \tag{2.19}$$

$$\boldsymbol{K} = \nabla^{\text{sym}} \boldsymbol{\varphi} \quad = K_{\alpha\beta}\ \boldsymbol{e}_\alpha \otimes \boldsymbol{e}_\beta \tag{2.20}$$

$$\boldsymbol{g} = \nabla w + \boldsymbol{\varphi} = g_\alpha \quad \boldsymbol{e}_\alpha \tag{2.21}$$

The tensors $\boldsymbol{G}$, $\boldsymbol{K}$, and $\boldsymbol{g}$ represent the in-plane strains, the curvature changes, and the transverse shear strains. $\boldsymbol{G}$ and $\boldsymbol{K}$ are second-order tensors, $\boldsymbol{g}$ is a first-order tensor. For the deflection gradient $\nabla w = \nabla^{\text{sym}} w$ holds true. Corresponding components of presented deformation measures are visualized in the Fig. 2.2 at and infinitesimal surface element. A purely qualitative presentation is chosen. Despite of the application of the mathematical rotation vector $\boldsymbol{\varphi}$, physically reasonable deformations are visualized.

2.3 Kinetics

Forces acting on the surface continuum represent the continuously distributed force densities. These arise from tangential stresses $\boldsymbol{s}$, moments $\boldsymbol{m}$ and transversal loads p. Herein, $\boldsymbol{f} = \boldsymbol{s} + p\boldsymbol{n}$ with the unit $^{\mathrm{N}}/_{\mathrm{m}^2}$ and $\boldsymbol{m} = -m_2\boldsymbol{e}_1 + m_1\boldsymbol{e}_2$ with the unit Nm hold. Analogous to the Cauchy theorem, the boundary measures are defined by these forces and moments acting at the surface.

$$\boldsymbol{n}_\nu = \lim_{\Delta L \to 0} \frac{\Delta \boldsymbol{s}}{\Delta L} \qquad \boldsymbol{m}_\nu = \lim_{\Delta L \to 0} \frac{\Delta(\boldsymbol{m} \times \boldsymbol{n})}{\Delta L} \qquad q_\nu = \lim_{\Delta L \to 0} \frac{\Delta p}{\Delta L} \tag{2.22}$$

The orientation of the cutting edges is described by the normals $\boldsymbol{n}$ und $\boldsymbol{\nu}$. The following applies to the edges with normal $\boldsymbol{n}$.

$$\boldsymbol{n} \cdot \boldsymbol{N} = \boldsymbol{o} \qquad \boldsymbol{n} \cdot \boldsymbol{L} = \boldsymbol{o} \qquad \boldsymbol{n} \cdot \boldsymbol{q} = 0 \tag{2.23}$$

By contrast, with the normal $\boldsymbol{\nu}$ pointing along the plane coordinate directions, the boundary loads are as follows.

$$\boldsymbol{\nu} \cdot \boldsymbol{N} = \boldsymbol{n}_\nu \qquad \boldsymbol{\nu} \cdot \boldsymbol{L} = \boldsymbol{m}_\nu \qquad \boldsymbol{\nu} \cdot \boldsymbol{q} = q_\nu \tag{2.24}$$

Here, the vectors and the scalar of the right hand sides indicate the boundary forces of the membrane state $\boldsymbol{n}_\nu$, the bending state $\boldsymbol{m}_\nu$ and of the transverse shear state q_ν. As with the Cauchy lemma, the cutting measures at opposite boundaries are the same, but contrary.

$$\boldsymbol{n}_\nu(-\boldsymbol{\nu}) = -\boldsymbol{n}_\nu(\boldsymbol{\nu}) \qquad \boldsymbol{m}_\nu(-\boldsymbol{\nu}) = -\boldsymbol{m}_\nu(\boldsymbol{\nu}) \qquad q_\nu(-\boldsymbol{\nu}) = -q_\nu(\boldsymbol{\nu}) \tag{2.25}$$

From Eqs. (2.23) and (2.24) tensors result for the so called stress resultants. Here, $\boldsymbol{N}$ is the membrane force tensor, $\boldsymbol{L}$ is the polar tensor of moments, and $\boldsymbol{q}$ is the transversal shear force vector. The tensors are constructed as follows.

$$\boldsymbol{N} = N_{\alpha\beta}\, \boldsymbol{e}_\alpha \otimes \boldsymbol{e}_\beta \qquad \boldsymbol{N} \in \mathbb{S}\mathrm{ym} \tag{2.26}$$

$$\boldsymbol{L} = M_{\alpha\beta}\, \boldsymbol{e}_\alpha \otimes \boldsymbol{e}_\beta \qquad \boldsymbol{L} \in \mathbb{S}\mathrm{ym} \tag{2.27}$$

$$\boldsymbol{q} = q_\alpha \quad \boldsymbol{e}_\alpha \tag{2.28}$$

The individual components are visualized in Fig. 2.3. Physically reasonable quantities are plotted. For the sake of completeness, the axial tensor of moments $\boldsymbol{M} = -\boldsymbol{L} \times \boldsymbol{n}$ is given, cf. [5], often found in literature.

$$\boldsymbol{M} = M_{11}\, \boldsymbol{e}_1 \otimes \boldsymbol{e}_2 + M_{21}\, \boldsymbol{e}_2 \otimes \boldsymbol{e}_2 - M_{22}\, \boldsymbol{e}_2 \otimes \boldsymbol{e}_1 - M_{12}\, \boldsymbol{e}_1 \otimes \boldsymbol{e}_1 \tag{2.29}$$

In addition to the introduced surface forces and moments, terms of inertia have to be introduced separately in the context of a surface continuum. An axiomatic

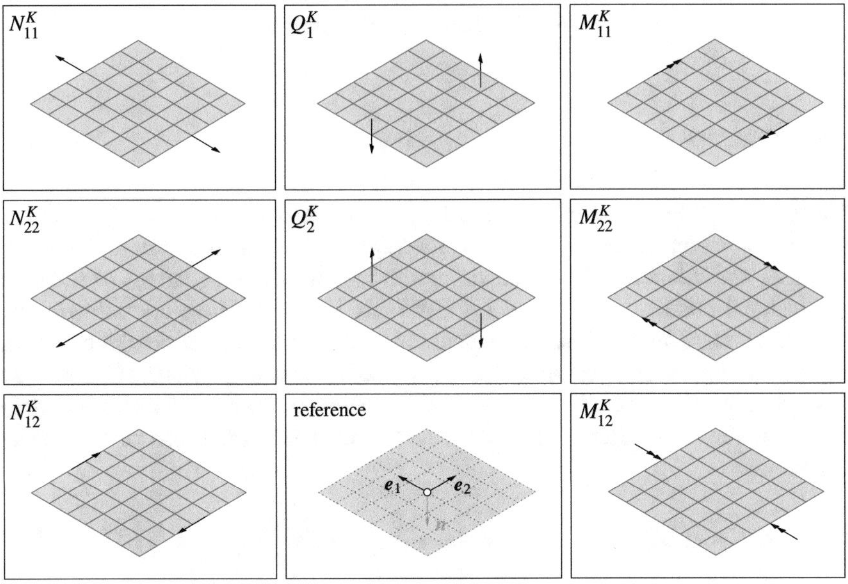

Fig. 2.3 Dual kinetic measures visualized at an infinitesimal surface element

introduction of these terms is not possible, contrary to the above statements on the force densities. However, following Zhilin [11], inertia terms for translations are $\boldsymbol{J}_{\mathrm{T}}$ and rotations $\boldsymbol{J}_{\mathrm{R}}$ can be introduced. These the terms of intertia are based on geometric measures in transverse direction. Since the surface continuum discussed here does not possess a thickness h at all, these quantities must be derived from the three dimensional continuum.

$$\boldsymbol{J}_{\mathrm{T}} = \int_{-h/2}^{+h/2} X_3 \mathrm{d}X_3 \boldsymbol{P} \times \boldsymbol{n} \tag{2.30}$$

$$\boldsymbol{J}_{\mathrm{R}} = \int_{-h/2}^{+h/2} X_3^2 \mathrm{d}X_3 \boldsymbol{P} \tag{2.31}$$

An exhaustive treatment of this problem can be found in [1, 11]. For what follows, it is beneficial to introduce a surface related mass density ρ, which can be determined as follows.

$$\rho = \int_{-h/2}^{+h/2} \rho_{\circledast} \mathrm{d}X_3 \tag{2.32}$$

Herein $\rho_{\circledast}$ is the mass density of the three dimensional continuum.

2.4 Balances

The mass balance is assumed to be fulfilled with regard to the classical continuum mechanics. In the context of the above statements, the balances are thus reduced to forces and moments. By transforming the Cauchy–Euler equations of the three dimensional continuum to the stress resultants introduced here and allowing for independent moments, the balance equations for the planar surface continuum result. These resulting Euler equations of the surface continuum are given here in a compact form. The division of the balance of forces into a membrane and a transverse shear state is omitted here. The local forms of these balance equations can be represented as follows.

$$\nabla \cdot (\boldsymbol{N} + \boldsymbol{q} \otimes \boldsymbol{n}) \qquad + \boldsymbol{s} + p\boldsymbol{n} = \rho \frac{\partial}{\partial t} \left[\dot{\boldsymbol{v}} + \dot{w}\boldsymbol{n} + \boldsymbol{J}_{\mathrm{T}}^{\top} \cdot (\boldsymbol{n} \times \dot{\boldsymbol{\varphi}})\right] \tag{2.33}$$

$$\nabla \cdot (-\boldsymbol{L} \times \boldsymbol{n}) + \boldsymbol{q} \times \boldsymbol{n} + \boldsymbol{m} \qquad = \rho \frac{\partial}{\partial t} \left[\boldsymbol{J}_{\mathrm{T}} \cdot (\dot{\boldsymbol{v}} + \dot{w}\boldsymbol{n}) + \boldsymbol{J}_{\mathrm{R}} \cdot (\boldsymbol{n} \times \dot{\boldsymbol{\varphi}})\right] \tag{2.34}$$

At this point it becomes obvious that the force tensor $\boldsymbol{F} = \boldsymbol{N} + \boldsymbol{q} \otimes \boldsymbol{n}$ is not symmetric. The balance of moments is considered separately contrary to the case of the Cauchy continuum. The coupling of both equations takes place via the vector invariant of the force tensor, which can be specified here with $\boldsymbol{F} \boxtimes \boldsymbol{1} = \boldsymbol{q} \times \boldsymbol{n}$, cf. p. 108 in Appendix A.

2.5 Initial and Boundary Conditions

In order to solve the field equations introduced above, boundary conditions are required for the considered surface continuum.

$$\partial\mathfrak{S} = \partial\mathfrak{S}_{\mathrm{D}} \cup \partial\mathfrak{S}_{\mathrm{N}} \qquad \partial\mathfrak{S}_{\mathrm{D}} \cap \partial\mathfrak{S}_{\mathrm{N}} = \emptyset \tag{2.35}$$

Subsequently, the conditions at these boundaries are specified. Quantities with an indexed * indicate prescribed measures.

2.5.1 Dirichlet Boundary Conditions

The Dirichlet boundary conditions are constraints in the form of given translations and rotations.

$$\begin{aligned} \boldsymbol{v}(\boldsymbol{r}_0) &= \boldsymbol{v}^\star(\boldsymbol{r}_0) \\ \boldsymbol{\varphi}(\boldsymbol{r}_0) &= \boldsymbol{\varphi}^\star(\boldsymbol{r}_0) \\ w(\boldsymbol{r}_0) &= w^\star(\boldsymbol{r}_0) \end{aligned} \qquad \forall\, \boldsymbol{r}_0 \in \partial\mathfrak{S}_\mathrm{D} \tag{2.36}$$

Homogeneous Dirichlet boundary conditions may also be prescribed.

$$\begin{aligned} \boldsymbol{v}(\boldsymbol{r}_0) &= \mathbf{o} \\ \boldsymbol{\varphi}(\boldsymbol{r}_0) &= \mathbf{o} \\ w(\boldsymbol{r}_0) &= 0 \end{aligned} \qquad \forall\, \boldsymbol{r}_0 \in \partial\mathfrak{S}_0 \tag{2.37}$$

2.5.2 Neumann Boundary Conditions

The Neumann boundary conditions relate forces and moments that can act as loads on the boundary of the surface continuum with the stress resultants.

$$\boldsymbol{\nu} \cdot \boldsymbol{N} = \boldsymbol{n}_\nu^\star \qquad \boldsymbol{\nu} \cdot \boldsymbol{L} = \boldsymbol{m}_\nu^\star \qquad \boldsymbol{\nu} \cdot \boldsymbol{q} = q_\nu^\star \qquad \forall\, \boldsymbol{r}_0 \in \partial\mathfrak{S}_\mathrm{N} \tag{2.38}$$

2.5.3 Initial Conditions

The initial state at time $t = t_0$ is given by the translations and rotations and their velocity fields.

$$\begin{aligned} \boldsymbol{v}(\boldsymbol{r}_0, t_0) &= \boldsymbol{v}(\boldsymbol{r}_0) & \dot{\boldsymbol{v}}(\boldsymbol{r}_0, t_0) &= \dot{\boldsymbol{v}}(\boldsymbol{r}_0) \\ \boldsymbol{\varphi}(\boldsymbol{r}_0, t_0) &= \boldsymbol{\varphi}(\boldsymbol{r}_0) & \dot{\boldsymbol{\varphi}}(\boldsymbol{r}_0, t_0) &= \dot{\boldsymbol{\varphi}}(\boldsymbol{r}_0) \\ w(\boldsymbol{r}_0, t_0) &= w(\boldsymbol{r}_0) & \dot{w}(\boldsymbol{r}_0, t_0) &= \dot{w}(\boldsymbol{r}_0) \end{aligned} \qquad \forall\, \boldsymbol{r}_0 \in \partial\mathfrak{S}_\mathrm{D} \tag{2.39}$$

2.6 Constitutive Relations

In context of present work we reduce ourselves to simple elastic materials, i.e. kinetic measures are determined by the actual state of deformation solely and therefore depend on the first gradient of the kinematics in maximum [6].

$$N = \mathcal{F}_1(\boldsymbol{G}) \tag{2.40}$$
$$L = \mathcal{F}_2(\boldsymbol{K}) \tag{2.41}$$
$$q = \mathcal{F}_3(\boldsymbol{g}) \tag{2.42}$$

Herein $\mathcal{F}_i\,\forall\, i \in \{1, 2, 3\}$ are elastic material laws. Only the symmetric parts of the kinematic gradients are considered, cf. Eqs. (2.19)–(2.21). In the case of small deformations, we can linearize the functions $\mathcal{F}_i$. Furthermore the decoupling of the membrane, bending, and transverse shear state holds true. This results in the following linear mappings whereby linear functions between the second-order tensors are represented by fourth order tensors while between first order tensors this is done by a second-order tensor.

$$\boldsymbol{N} = \boldsymbol{\mathcal{A}} : \boldsymbol{G} \tag{2.43}$$
$$\boldsymbol{L} = \boldsymbol{\mathcal{D}} : \boldsymbol{K} \tag{2.44}$$
$$\boldsymbol{q} = \boldsymbol{Z} \cdot \boldsymbol{g} \tag{2.45}$$

$\boldsymbol{\mathcal{A}}$ is the fourth-order positive-definite tensor of linear, in-plane, elastic stiffness, $\boldsymbol{\mathcal{D}}$ is the fourth-order positive-definite tensor of linear, out-of-plane elastic stiffnesses and $\boldsymbol{Z}$ is the second-order positive-definite tensor of linear, transversal, elastic shear stiffnesses. In contrast to the constitutive tensor of classical continuum mechanics, $\boldsymbol{\mathcal{A}}$ and $\boldsymbol{\mathcal{D}}$ have only 2^4 components, $\boldsymbol{Z}$ only 2^2. The following symmetry properties apply to $\boldsymbol{\mathcal{H}} = \{\boldsymbol{\mathcal{A}}, \boldsymbol{\mathcal{D}}\}$ [3, 4]

$$\begin{aligned}
\boldsymbol{A} : \boldsymbol{\mathcal{H}} : \boldsymbol{B} &= \boldsymbol{B} : \boldsymbol{\mathcal{H}} : \boldsymbol{A} && H_{\alpha\beta\gamma\delta} = H_{\gamma\delta\alpha\beta} && \text{(main symmetry)}\\
\boldsymbol{A} : \boldsymbol{\mathcal{H}} &= \boldsymbol{A}^{\top} : \boldsymbol{\mathcal{H}} && H_{\alpha\beta\gamma\delta} = H_{\beta\alpha\gamma\delta} && \text{(left subsymmetry)}\\
\boldsymbol{\mathcal{H}} : \boldsymbol{A} &= \boldsymbol{\mathcal{H}} : \boldsymbol{A}^{\top} && H_{\alpha\beta\gamma\delta} = H_{\alpha\beta\delta\gamma} && \text{(right subsymmetry)}
\end{aligned}$$

and $\boldsymbol{Z}$ [3].

$$\boldsymbol{Z} \cdot \boldsymbol{b} = \boldsymbol{b} \cdot \boldsymbol{Z} \qquad Z_{\alpha\beta} = Z_{\beta\alpha} \qquad \text{(symmetry)}$$

Herein $\boldsymbol{A} = A_{\alpha\beta}\boldsymbol{e}_\alpha \otimes \boldsymbol{e}_\beta$, $\boldsymbol{B} = B_{\alpha\beta}\boldsymbol{e}_\alpha \otimes \boldsymbol{e}_\beta$, and $\boldsymbol{b} = b_\alpha \boldsymbol{e}_\alpha$ are chosen arbitrary. For the stiffness tensors, projector representations can be found through spectral decompositions.

$$\boldsymbol{\mathcal{A}} = \sum_{\alpha=1}^{2n} \lambda_\alpha^{\mathcal{A}} \boldsymbol{\mathcal{P}}_\alpha^{\mathfrak{S}} \qquad \boldsymbol{\mathcal{D}} = \sum_{\alpha=1}^{2n} \lambda_\alpha^{\mathcal{D}} \boldsymbol{\mathcal{P}}_\alpha^{\mathfrak{S}} \qquad \boldsymbol{Z} = \sum_{\alpha=1}^{n} \lambda_\alpha^{Z} \boldsymbol{P}_\alpha^{\mathfrak{S}} \tag{2.46}$$

In the general case, $n = 2$ holds, so that $\boldsymbol{\mathcal{A}}$ and $\boldsymbol{\mathcal{D}}$ have $2n \leq 4$ distinct real eigenvalues $\lambda_\alpha^{\mathcal{A}}$, $\lambda_\alpha^{\mathcal{D}}$, while $\boldsymbol{Z}$ has $n \leq 2$ distinct real eigenvalues λ_α^{Z}. The properties of the fourth-order projectors $\boldsymbol{\mathcal{P}}^{\mathfrak{S}}$ are as follows.

$$\mathcal{P}^{\mathfrak{S}}_{\alpha} : \mathcal{P}^{\mathfrak{S}}_{\beta} = \begin{cases} \mathcal{P}^{\mathfrak{S}}_{\alpha} & \text{if } \alpha = \beta \\ \mathcal{O} & \text{if } \alpha \neq \beta \end{cases} \qquad \sum_{\alpha=1}^{n} \mathcal{P}^{\mathfrak{S}}_{\alpha} = \mathcal{P}^{\text{sym}} \tag{2.47}$$

Herein, $\mathcal{P}^{\text{sym}}$ is the symmetric part of the fourth-order identity of the surface continuum. The second-order projector $\boldsymbol{P}^{\mathfrak{S}}_{\alpha}$ has the following characteristics.

$$\boldsymbol{P}^{\mathfrak{S}}_{\alpha} \cdot \boldsymbol{P}^{\mathfrak{S}}_{\beta} = \begin{cases} \boldsymbol{P}^{\mathfrak{S}}_{\alpha} & \text{if } \alpha = \beta \\ \boldsymbol{0} & \text{if } \alpha \neq \beta \end{cases} \qquad \sum_{\alpha=1}^{n} \boldsymbol{P}^{\mathfrak{S}}_{\alpha} = \boldsymbol{P} \tag{2.48}$$

The isotropy assumption leads to $n = 1$. The projectors for the stiffness tensors are as follows.

$$\mathcal{P}^{\mathfrak{S}}_{1} = \frac{1}{2} \boldsymbol{P} \otimes \boldsymbol{P} \tag{2.49}$$

$$\mathcal{P}^{\mathfrak{S}}_{2} = \mathcal{P}^{\text{sym}} - \mathcal{P}^{\mathfrak{S}}_{1} \tag{2.50}$$

$$\boldsymbol{P}^{\mathfrak{S}}_{1} = \boldsymbol{e}_{\alpha} \otimes \boldsymbol{e}_{\alpha} \tag{2.51}$$

The second-order projector $\boldsymbol{P}^{\mathfrak{S}}_{1}$ coincides with the first metric tensor of the surface $\boldsymbol{P}$ in the isotropic case. The Eq. (2.46) can now be specified using these projectors.

$$\mathcal{A} = \lambda^{\mathcal{A}}_{1} \mathcal{P}^{\mathfrak{S}}_{1} + \lambda^{\mathcal{A}}_{2} \mathcal{P}^{\mathfrak{S}}_{2} \qquad \mathcal{D} = \lambda^{\mathcal{D}}_{1} \mathcal{P}^{\mathfrak{S}}_{1} + \lambda^{\mathcal{D}}_{2} \mathcal{P}^{\mathfrak{S}}_{2} \qquad \boldsymbol{Z} = \lambda^{Z}_{1} \boldsymbol{P} \tag{2.52}$$

The isotropic eigenvalues are as follows.

$$\lambda^{\mathcal{A}}_{1} = \frac{Yh}{1-\nu} = 2Bh \qquad \lambda^{\mathcal{A}}_{2} = \frac{Yh}{1+\nu} = 2Gh \tag{2.53}$$

$$\lambda^{\mathcal{D}}_{1} = \frac{Yh^3}{12(1-\nu)} = 2B\frac{h^3}{12} \qquad \lambda^{\mathcal{D}}_{2} = \frac{Yh^3}{12(1+\nu)} = 2G\frac{h^3}{12} \tag{2.54}$$

$$\lambda^{Z}_{1} = \frac{\kappa\, Y h}{2(1+\nu)} = \kappa\, G\, h \tag{2.55}$$

Herein B is the bulk modulus of the surface continuum and G is the shear modulus. In λ^{Z}_{1}, κ is a correction factor to account for transverse shear deformations. When considering individual layers, this is restricted to $0 < \kappa \leq 1$ [10]. It serves to adjust the shear energy contribution. The eigenvalues $\lambda^{\mathcal{A}}_{1}$ and $\lambda^{\mathcal{D}}_{1}$ represent resistances against changes in area, where $\lambda^{\mathcal{A}}_{2}$, $\lambda^{\mathcal{D}}_{2}$, and λ^{Z}_{1} indicate resistances to changes in shape. A more detailed account of the structure of the projectors describing the elasticity of the surface continuum and explanations of the strategy shown here can be found in [3] or [2], respectively. As an alternative to the derivation for the Hooke's law, the stiffness tensors can also be derived through the postulate of an elastic potential. Such an approach is attached to Appendix B, where the elastic potential is given in explicit form in the context of present considerations.

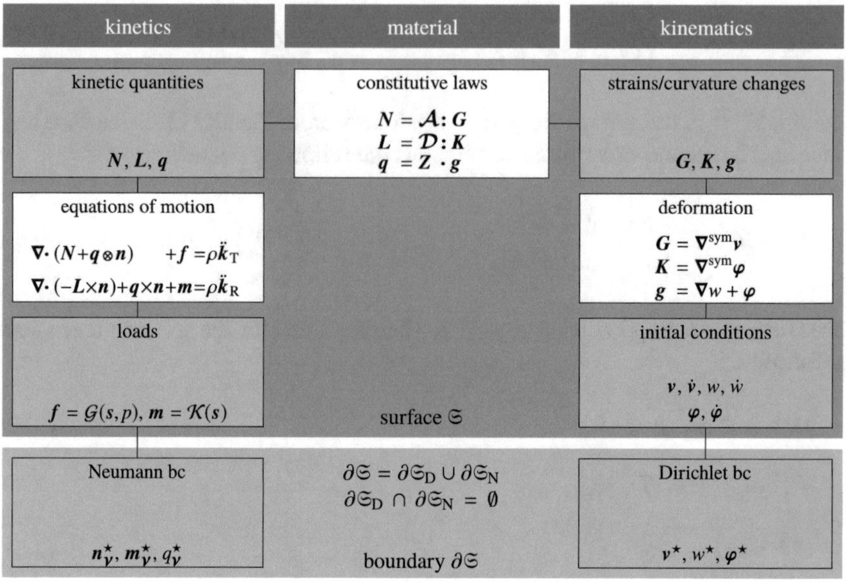

Fig. 2.4 Structural scheme of the direct formulated theory for planar surface continua

Based on the projector representation, the constitutive equations can be formulated by the aid of dilatoric and deviatorc deformation portions, cf. p. 109 in Appendix A.

$$\boldsymbol{N} = 2Bh \quad \boldsymbol{G}^{\text{dil}} + 2Gh \quad \boldsymbol{G}^{\text{dev}} \tag{2.56}$$

$$\boldsymbol{L} = 2B\frac{h^3}{12}\boldsymbol{K}^{\text{dil}} + 2G\frac{h^3}{12}\boldsymbol{K}^{\text{dev}} \tag{2.57}$$

$$\boldsymbol{q} = 2G\frac{\kappa h}{2}\ \boldsymbol{g} \tag{2.58}$$

Correlations to classical engineering parameters, e.g. Young's modulus Y and Poisson's ratio ν are as follows.

$$B = \frac{Y}{2(1-\nu)} \qquad\qquad G = \frac{Y}{2(1+\nu)} \tag{2.59}$$

2.7 Formalization

At this point a Tonti diagram [8] is generated for the above equations and their correlations. This classification diagram is used for the explicit, graphical processing of the initial boundary value problem. Particularly clear is the compact description, possible by the direct tensor notation. For the sake of clarity, the right hand sides of the equations of motion (2.33) and (2.34) are abbreviated as follows. The translatoric degrees of freedom are summarized in $\boldsymbol{a}$, cf. Eq. (2.12).

$$\rho \ddot{\boldsymbol{k}}_{\mathrm{T}} = \rho \left(\ddot{\boldsymbol{a}} + \boldsymbol{J}_{\mathrm{T}}^{\top} \cdot \boldsymbol{n} \times \ddot{\boldsymbol{\varphi}} \right) \tag{2.60}$$

$$\rho \ddot{\boldsymbol{k}}_{\mathrm{R}} = \rho \left(\boldsymbol{J}_{\mathrm{T}} \cdot \ddot{\boldsymbol{a}} + \boldsymbol{J}_{\mathrm{R}} \cdot \boldsymbol{n} \times \ddot{\boldsymbol{\varphi}} \right) \tag{2.61}$$

The result of this structuring is given in Fig. 2.4. Obvious here is the right column with the kinematic variables and the dynamic variables in the left column. In the middle are the linear mappings [9]. Regarding the loads, $\mathcal{G}$ and $\mathcal{F}$ stand for arbitrary and independent functions.

References

1. Aßmus M (2018) Global structural analysis at photovoltaic modules: theory, numerics, application (in German). Dissertation, Otto von Guericke University Magdeburg
2. Aßmus M, Eisenträger J, Altenbach H (2017) On isotropic linear elastic material laws for directed planes. In: Proceedings of the 11th international conference on shell structures: theory and applications (SSTA 2017), Gdańsk, Poland, pp 57–60. https://www.taylorfrancis.com/books/e/9781351680486/chapters/10.1201%2F9781315166605-7
3. Aßmus M, Eisenträger J, Altenbach H (2017) Projector representation of isotropic linear elastic material laws for directed surfaces. Zeitschrift für Angewandte Mathematik und Mechanik 97(-):1–10. https://doi.org/10.1002/zamm.201700122
4. Bertram A, Glüge R (2015) Solid mechanics: theory, modeling, and problems. Springer, Cham. https://doi.org/10.1007/978-3-319-19566-7
5. Libai A, Simmonds JG (1983) Nonlinear elastic shell theory. Adv Appl Mech 23:271–371. https://doi.org/10.1016/S0065-2156(08)70245-X
6. Noll W (1958) A mathematical theory of the mechanical behavior of continuous media. Arch Rat Mech Anal 2(1):197–226. https://doi.org/10.1007/BF00277929
7. Pal'mov VA (1964) Fundamental equations of the theory of asymmetric elasticity. J Appl Math Mech 28(3):496–505. https://doi.org/10.1016/0021-8928(64)90092-9
8. Tonti E (1972) On the mathematical structure of a large class of physical theories. Accademia Nazionale Dei Lincei 52(1):48–56
9. Tonti E (2013) The mathematical structure of classical and relativistic physics - a general classification diagram. Birkhäuser, Basel. https://doi.org/10.1007/978-1-4614-7422-7
10. Vlachoutsis S (1992) Shear correction factors for plates and shells. Int J Numer Methods Eng 33(7):1537–1552. https://doi.org/10.1002/nme.1620330712
11. Zhilin PA (2006) Applied mechanics - foundations of shells theory (in russian). Publisher of the Polytechnic University, St. Petersburg. http://mp.ipme.ru/Zhilin/Zhilin_New/pdf/Zhilin_Shell_Book.pdf

Chapter 3
Multilayered Surface Continua

3.1 The Concept of Multiple Layers

The computation of composite structures, as introduced in Chap. 1, requires the extension of the presented concept of the surface continuum to multiple layers. This is especially true when the physical layer thicknesses differ widely and the mechanical properties of the layer materials are strongly divergent. This is the case with Anti-Sandwiches. In this sense, the present chapter introduces a so-called layer-wise theory. Each layer is considered individually, whereby the coupling is realized via kinematic constraints. However, apart from the consideration of the individual mid surfaces, this procedure also requires the consideration of interfaces of the physical structure. The concept follows the work of Naumenko and Eremeyev [5], for which the term eXtended layerwise theory (XLWT) has become established [3].

3.2 Assumptions and Restrictions

In what follows, the restriction to the coupling of three individual layers holds. The following assumptions and restrictions should be made.

- All surfaces are plane-parallel in the reference placement.
- All surfaces are at a constant distance from each other for all admissible deformations.
- Thus, the normal per layer does not undergo elongation. It is inextensible.
- This implies that the thickness strains E_{33} per layer are identical to zero.
- The mid surfaces are coupled via Mindlin's straight line hypothesis at the interfaces between the layers.
- This implies that we need to consider separately the transverse shears $E_{\alpha 3}$ as introduced in Chap. 2.

M. Aßmus, *Structural Mechanics of Anti-Sandwiches*,
SpringerBriefs in Continuum Mechanics,
https://doi.org/10.1007/978-3-030-04354-4_3

- At the interfaces between the layers, no damage occurs, the layers are thus perfectly bonded for any deformation states.
- This implies that a slipping of the individual layers at the interfaces is prevented. Delamination is excluded.

3.3 Global Coordinate System and Transverse Boundaries

With the introduction of the composition of three layers, which initially have a layer-specific coordinate system, a global coordinate system for the composite is to be determined. It has its origin in the middle of the core layer and runs to the two outer faces of the composite structure. Thus, for the X_3 coordinate

$$-h^t - {}^{h^c}\!/_2 \leq X_3 \leq {}^{h^c}\!/_2 + h^b \tag{3.1}$$

applies. The composition of the layers is visualized in Fig. 3.1. For the three layers to which the calculations are related, the indices t, c and b are introduced. Here t stands for the top layer, c for core layer, and b for back layer. In addition to the layer mid surfaces $\mathfrak{S}^K \ \forall K \in \{t, c, b\}$ there are the interfaces $\mathfrak{I}^K \ \forall K \in \{t, b\}$ and the two outer faces $\mathfrak{V}$ and $\mathfrak{R}$ are visible. The distance of a surface $\mathfrak{S}^K$ to the interfaces is

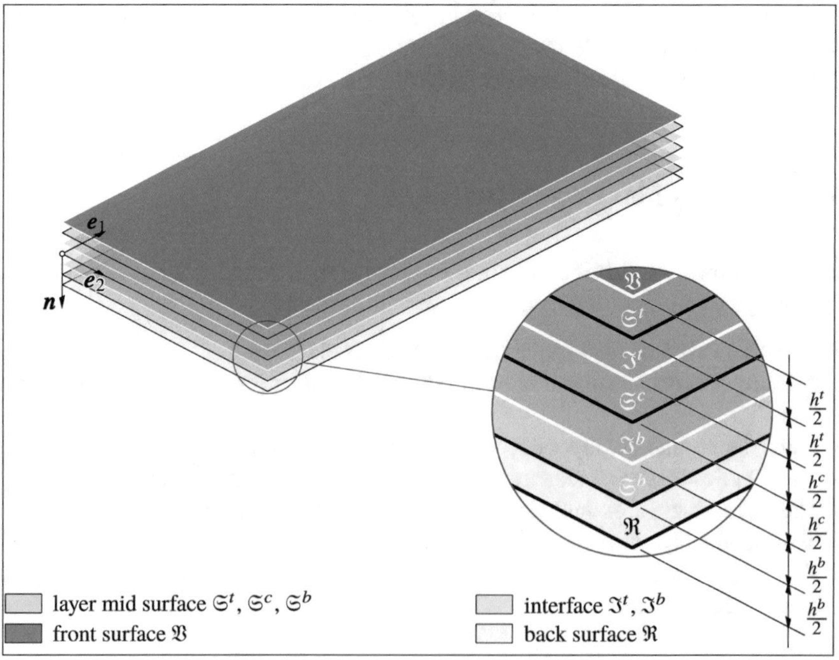

Fig. 3.1 Structure with multiple plane surface continua and physical boundaries

$\mp h^K/2$. The newly introduced, global coordinate system thus has its origin at the core layers surface $\mathfrak{S}^c$.

The quantities that were previously accounted for balancing must now be determined according to the newly introduced coordinate origin, taking into account the new integration limits. The integration limits for quantities of top layer (t), core layer (c), and back layer (b) can be specified as follows.

$$\Box^t : \Big|_{-\frac{h^c}{2}-h^t}^{-\frac{h^c}{2}} \qquad \Box^c : \Big|_{-\frac{h^c}{2}}^{+\frac{h^c}{2}} \qquad \Box^b : \Big|_{+\frac{h^c}{2}}^{+\frac{h^c}{2}+h^b} \tag{3.2}$$

Herein, $\Box^K \;\forall K \in \{t, c, b\}$ indicates the variable to be integrated. Due to the orientation of the coordinate system, the lower bounds of integration are in the direction of the top layer ($-\boldsymbol{n}$) and the upper bounds of integration are in the direction of the back layer ($+\boldsymbol{n}$).

3.4 Coupling Conditions

The preceding section describes the geometric position of the three layers. At this point, kinematic constraints for coupling the three layers are to be introduced. First, the deflection in all three layers should be identical.

$$w^t = w^c = w^b = w \qquad \forall\, \boldsymbol{r}^K = X^K_\alpha\, \boldsymbol{e}_\alpha \tag{3.3}$$

The consequence of this restriction is that no changes in the thickness of the physical structure can occur.

$$\frac{\partial h^K}{\partial X_1} = \frac{\partial h^K}{\partial X_2} = 0 \tag{3.4}$$

This also eliminates normal strains in the thickness direction.

$$\frac{\partial w}{\partial X_3^K} = 0 \qquad \Rightarrow \qquad E_{33} = 0 \tag{3.5}$$

The deflections are thus equal $w = w^K \;\forall K \in \{t, c, b\}$ for every material point $\boldsymbol{r} \in \mathfrak{S}^K \;\forall K \in \{t, c, b\}$ of each surface. The number of degrees of freedom of the composite can be reduced from 15 to 13.

Another restriction concerns the connection of the layers at the interfaces $\mathfrak{I}^K \;\forall K \in \{t, b\}$ with normal $\pm\boldsymbol{n}$. All layers should be firmly connected at these interfaces. Delamination is thus excluded as previously mentioned. The following kinematic constraints can thus be introduced since the normal hypothesis related to Mindlin's first-order shear deformation theory is valid.

$$\boldsymbol{v}^t + \frac{h^t}{2}\,\boldsymbol{\psi}^t = \boldsymbol{v}^c - \frac{h^c}{2}\,\boldsymbol{\psi}^c \qquad \forall\,\mathfrak{I}^t \tag{3.6}$$

$$\boldsymbol{v}^b - \frac{h^b}{2}\,\boldsymbol{\psi}^b = \boldsymbol{v}^c + \frac{h^c}{2}\,\boldsymbol{\psi}^c \qquad \forall\,\mathfrak{I}^b \tag{3.7}$$

The above restrictions lead to a further reduction in the number of degrees of freedom. By this formulation, all quantities of the core layer ($\boldsymbol{v}^c = v_1^c\boldsymbol{e}_1 + v_2^c\boldsymbol{e}_2$ and $\boldsymbol{\psi}^c = -\varphi_2^c\boldsymbol{e}_1 + \varphi_1^c\boldsymbol{e}_2$ or $\boldsymbol{\varphi}^c = \varphi_1^c\boldsymbol{e}_1 + \varphi_2^c\boldsymbol{e}_2$ by $\boldsymbol{\varphi}^c = \boldsymbol{\psi}^c \times \boldsymbol{n}$, resp.), can be substituted by quantities of the skin layers.

$$\boldsymbol{\psi}^c = \frac{1}{h^c}\left[\boldsymbol{v}^b - \boldsymbol{v}^t - \frac{h^b}{2}\boldsymbol{\psi}^b - \frac{h^t}{2}\boldsymbol{\psi}^t\right] \tag{3.8}$$

$$\boldsymbol{v}^c = \frac{1}{2}\left[\boldsymbol{v}^t + \boldsymbol{v}^b + \frac{h^t}{2}\boldsymbol{\psi}^t - \frac{h^b}{2}\boldsymbol{\psi}^b\right] \tag{3.9}$$

The resulting number of degrees of freedom of the overall system thus drops to 9.

3.5 Quantities at Composite Level

In Chap. 2 the loads have been introduced at the surface $\mathfrak{S}$. However, at the physical structure, the loads act at a distance of $\mp h/2$ to the individual surface $\mathfrak{S}$. In the context of the structure introduced in Sect. 3.3, this concerns the top $\mathfrak{V}$ and the back surface $\mathfrak{R}$. The force load vector can be formulated as follows, taking into account the interactions that result at the interfaces ($\mathfrak{I}^t$, $\mathfrak{I}^b$). The back surface of the composite should remain unloaded for the sake of simplicity.

$$\boldsymbol{f}^K = \begin{cases} \left(q^t + p\right)\boldsymbol{n} + \boldsymbol{s}^t - \boldsymbol{s} & \text{if } K = t \\ \left(q^b - q^t\right)\boldsymbol{n} + \boldsymbol{s}^b - \boldsymbol{s}^t & \text{if } K = c \\ q^b\,\boldsymbol{n} - \boldsymbol{s}^b & \text{if } K = b \end{cases} \qquad \forall\,\boldsymbol{f}^K \in \mathfrak{S}^K \tag{3.10}$$

Herein, p is an orthogonal force acting at $\mathfrak{V}$, q^K are orthogonal forces at the interfaces, $\boldsymbol{s}^K$ are longitudinal forces, while $\boldsymbol{s}$ is acting at $\mathfrak{V}$ solely. All indexed measures refer to their corresponding surface. In the context of present work, moments acting at the composite should only result from acting tangential loads. The moments acting on the surface $\mathfrak{V}$ and those interacting with the other layers may be given for each layer as follows.

$$\boldsymbol{m}^K = \begin{cases} \frac{h^t}{2}\boldsymbol{n} \times \left(\boldsymbol{s} + \boldsymbol{s}^t\right) & \text{if } K = t \\ \frac{h^c}{2}\boldsymbol{n} \times \left(\boldsymbol{s}^t + \boldsymbol{s}^b\right) & \text{if } K = c \\ \frac{h^b}{2}\boldsymbol{n} \times \boldsymbol{s}^b & \text{if } K = b \end{cases} \qquad \forall\,\boldsymbol{m}^K \in \mathfrak{S}^K \tag{3.11}$$

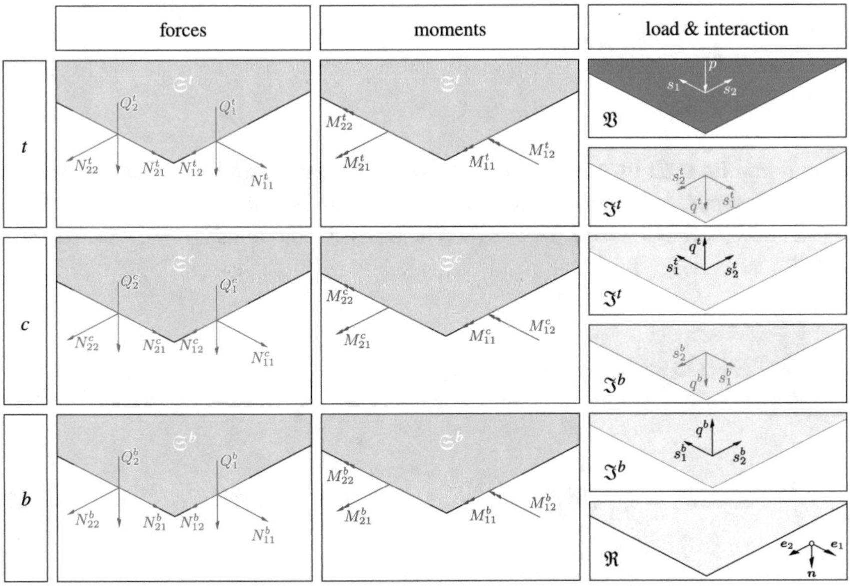

Fig. 3.2 Forces, moments, loads, and interactions at the individual layers

Forces, moments as well as loads and interaction quantities are visualized in Fig. 3.2.

The determination of the terms of inertia from Eqs. (2.30) and (2.31) requires increased attention. This results from the global position of the integration bounds determined in Eq. (3.2). First, the integration of the thickness coordinate for the translational inertia term should be made.

$$\int_{-\frac{h^c}{2}-h^t}^{-\frac{h^c}{2}} X_3 \, \mathrm{d}X_3 = \frac{1}{2}\left[\left(\frac{h^c}{2}\right)^2 - \left(-\frac{h^c}{2} - h^t\right)^2\right] = -\frac{h^t}{2}\left(h^c + h^t\right) = \alpha^t \quad (3.12)$$

$$\int_{-\frac{h^c}{2}}^{+\frac{h^c}{2}} X_3 \, \mathrm{d}X_3 = \left[\frac{(h^c)^2}{8} - \frac{(h^c)^2}{8}\right] = 0 = \alpha^c \quad (3.13)$$

$$\int_{+\frac{h^c}{2}}^{+\frac{h^c}{2}+h^b} X_3 \, \mathrm{d}X_3 = \frac{1}{2}\left[\left(\frac{h^c}{2} + h^b\right)^2 - \left(\frac{h^c}{2}\right)^2\right] = \frac{h^b}{2}\left(h^c + h^b\right) = \alpha^b \quad (3.14)$$

It introduces $\alpha^K \ \forall K \in \{t, c, b\}$ as shorthand for geometric relations in context of a compact notation. The tensors of the translational terms of inertia thus result as follows.

$$J_{\mathrm{T}}^{t} = \alpha^{t}\ \boldsymbol{P} \times \boldsymbol{n} \tag{3.15}$$

$$J_{\mathrm{T}}^{c} = \alpha^{c}\ \boldsymbol{0} \qquad = \boldsymbol{0} \tag{3.16}$$

$$J_{\mathrm{T}}^{b} = \alpha^{b}\ \boldsymbol{P} \times \boldsymbol{n} \tag{3.17}$$

Here it can be seen that the term of the translational tensor of inertial of the core layer disappears, but the proportions of the two skin layers do not.

The integration of the thickness coordinate for the rotational term of inertia results as follows.

$$\int_{-\frac{h^c}{2}-h^t}^{-\frac{h^c}{2}} X_3^2\,\mathrm{d}X_3 = h^t\left[\frac{1}{2}h^c h^t + \frac{1}{4}\left(h^c\right)^2 + \frac{1}{3}\left(h^t\right)^2\right] = \beta^t \tag{3.18}$$

$$\int_{-\frac{h^c}{2}}^{+\frac{h^c}{2}} X_3^2\,\mathrm{d}X_3 = \frac{1}{12}\left(h^c\right)^3 \qquad = \beta^c \tag{3.19}$$

$$\int_{+\frac{h^c}{2}}^{+\frac{h^c}{2}+h^b} X_3^2\,\mathrm{d}X_3 = h^b\left[\frac{1}{2}h^c h^b + \frac{1}{4}\left(h^c\right)^2 + \frac{1}{3}\left(h^b\right)^2\right] = \beta^b \tag{3.20}$$

The expressions $\beta^K\ \forall\, K \in \{t, c, b\}$ are also abbreviations for geometrical relationships. The tensors of the rotational inertia can thus be stated as follows.

$$J_{\mathrm{R}}^{t} = \beta^{t}\,\boldsymbol{P} \tag{3.21}$$

$$J_{\mathrm{R}}^{c} = \beta^{c}\,\boldsymbol{P} \tag{3.22}$$

$$J_{\mathrm{R}}^{b} = \beta^{b}\,\boldsymbol{P} \tag{3.23}$$

Since the transposed of the tensor of the translational inertia $\left[J_{\mathrm{T}}^{K}\right]^{\top}$ is also required in the force equilibrium equations of the individual layers, their structure should also be given for the sake of completeness.

$$\left[J_{\mathrm{T}}^{t}\right]^{\top} = \left[\frac{h^t}{2}\left(h^c + h^t\right)\right] \boldsymbol{P} \times \boldsymbol{n} = -\alpha^t\ \boldsymbol{P} \times \boldsymbol{n} \tag{3.24}$$

$$\left[J_{\mathrm{T}}^{c}\right]^{\top} = \boldsymbol{0} \qquad = -\alpha^c\ \boldsymbol{0} \tag{3.25}$$

$$\left[J_{\mathrm{T}}^{b}\right]^{\top} = -\left[\frac{h^b}{2}\left(h^c + h^b\right)\right] \boldsymbol{P} \times \boldsymbol{n} = -\alpha^b\ \boldsymbol{P} \times \boldsymbol{n} \tag{3.26}$$

Here, too, the terms of the core layer disappear completely.

3.6 Strong Formulation of Initial Boundary Value Problem

Due to the introduced restrictions, the strong forms of the differential equations of the balances can be specified. For this purpose, the force and moment balance are formulated layerwise while the terms for loads, interactions and inertia determined in this chapter are used. In doing so, the membrane force equilibrium is determined by projecting the force equation into the plane (Eq. (2.33) $\cdot\, \boldsymbol{P}$) and the transverse shear force equilibrium is generated by the out-of-plane projection of the force equilibrium (Eq. (2.33) $\cdot\, \boldsymbol{n}$). Furthermore, the moment equilibrium is formulated in terms of polar vectors and tensors (Eq. (2.34)$\times\boldsymbol{n}$).

$$\nabla \cdot \boldsymbol{N}^t + \boldsymbol{s}^t - \boldsymbol{s} = \rho^t \left(\ddot{\boldsymbol{v}}^t + \alpha^t \ddot{\boldsymbol{\varphi}}^b\right) \tag{3.27}$$

$$\nabla \cdot \boldsymbol{N}^c + \boldsymbol{s}^b - \boldsymbol{s}^t = \rho^c \ddot{\boldsymbol{v}}^c \tag{3.28}$$

$$\nabla \cdot \boldsymbol{N}^b - \boldsymbol{s}^b = \rho^t \left(\ddot{\boldsymbol{v}}^b + \alpha^b \ddot{\boldsymbol{\varphi}}^b\right) \tag{3.29}$$

$$\nabla \cdot \boldsymbol{q}^t + q^t + q = \rho^t \ddot{w}^t \tag{3.30}$$

$$\nabla \cdot \boldsymbol{q}^c + q^b - q^t = \rho^c \ddot{w}^c \tag{3.31}$$

$$\nabla \cdot \boldsymbol{q}^b - q^b = \rho^b \ddot{w}^b \tag{3.32}$$

$$\nabla \cdot \boldsymbol{L}^t - \boldsymbol{q}^t + \frac{h^t}{2}\left(\boldsymbol{s} + \boldsymbol{s}^t\right) = \rho^t \left(\alpha^t \ddot{\boldsymbol{v}}^t + \beta^t \ddot{\boldsymbol{\varphi}}^t\right) \tag{3.33}$$

$$\nabla \cdot \boldsymbol{L}^c - \boldsymbol{q}^c + \frac{h^c}{2}\left(\boldsymbol{s}^t + \boldsymbol{s}^b\right) = \rho^c \beta^c \ddot{\boldsymbol{\varphi}}^c \tag{3.34}$$

$$\nabla \cdot \boldsymbol{L}^b - \boldsymbol{q}^b + \frac{h^b}{2}\boldsymbol{s}^b = \rho^b \left(\alpha^b \ddot{\boldsymbol{v}}^b + \beta^b \ddot{\boldsymbol{\varphi}}^b\right) \tag{3.35}$$

For these nine equations resulting for the entire composite, a solution strategy is now required. Closed-form solutions of this set of equations are not possible. Naumenko and Eremeyev [5] present an closed-form solution under severe constraints, such as the transition to equations of the shear-rigid theory for the skin layers. Without devoting itself to such simplification, a numerical strategy for solving problems that may be as general as possible is to be developed in present work.

3.7 On Shear Correction

An initially indefinite parameter is κ. The introduction of this shear correction factor $0 < \kappa \leq 1$ can influence the accuracy of the calculation positively in the presence of transverse shear deformations. It must be determined individually for each layer [4] and is a parameter for the regulation of the transverse shear energy fraction [6].

$$D_S^K = \kappa^K G^K h^K \tag{3.36}$$

The introduction of this factor becomes necessary because the transverse shear stress curve takes a parabolic course in the thickness direction, see illustration Fig. 3.3 on the left-hand side, and the shear stiffness of the examined structure does not tend towards infinity. Originally the idea came from Timoshenko, which stood before similar problems at beams in 1921 [8, 9]. However, due to the fact that the shear stiffness D_S^K takes into account only a mean value, since it is related to the mid surface of the structure, only a constant gradient is taken into account, see Fig. 3.3 in the center. This correction has already been introduced by Reissner. Reissner derived $\kappa = {}^5/_6$ while Mindlin specified the factor with $\kappa = {}^{\pi^2}/_{12}$. The difference in both results is due to the different approaches of both authors. While Reissner comes to conclusions about energy considerations in static deformation states, Mindlin extended the plate equations by terms of inertia and found the correction factor over eigenfrequency calculations. For practical calculations, the difference has little relevance.

Since each layer is taken into account individually in the present work, a single shear correction factor must also be found for each layer. For the following considerations, the complementary energy $W^\star = W(\boldsymbol{T})$ is used on the basis of the force approach. Considering a simple case for Q_2 with the transverse shear stress $T_{23}(X_3) = Q_2 S(X_3) I^{-1}$ with the parabolic function $S(X_3) = {}^1/_2[({}^h/_2)^2 - X_3^2]$, which represents the static moment (1. moment of area) of the cross-sectional portion relative to the width 1 with respect to the mid surface, and the moment of inertia $I = {}^{h^3}/_{12}$ of the plate cross-section with width 1 and height h, the following system of equations can be set up for a single layer [1, 2].

$$\begin{aligned} W^\star_{\text{real}} &= W^\star_{\text{const}} \\ \frac{1}{2}\int_{-\frac{h}{2}}^{+\frac{h}{2}} \frac{T_{23}^2(X_3)}{G}\,\mathrm{d}X_3 &= \frac{1}{2}\int_{-\frac{h}{2}}^{+\frac{h}{2}} \frac{Q_2^2}{\kappa G h^2}\,\mathrm{d}X_3 \\ \frac{1}{2}\frac{Q_2^2}{Gh}\frac{6}{5} &= \frac{1}{2}\frac{Q_2^2}{\kappa G h} \end{aligned} \tag{3.37}$$

Thus, the energy-equivalent complementary energy is given by $W^\star_{\text{const}} = \frac{5}{6}\,W^\star_{\text{real}}$. This shear correction is denoted by $\kappa = \frac{5}{6}$ in the classical literature and in the context of Eq. (3.36). The use of the terms Q_1 and T_{13} for determining the complementary energy for analogue applications in Eq. (3.37) leads to identical shear correction factors for isotropic materials. Since in the present considerations any physical orientation dependence is excluded, $\kappa = \kappa_1 = \kappa_2$ holds true. In [7, 10] it has been determined on the basis of experimental investigations, that the shear correction factor $\kappa = 1$ shows the best matches between experiment and computation for the present problem. In context of the above derivations, this choice seems questionable at first. However, taking into account that the relevant shear-soft core layer is very thin, it can be seen that the static moment loses its influence. Although the function always remains parabolic, the amplitude is proportional to the structural thickness. With vanishing

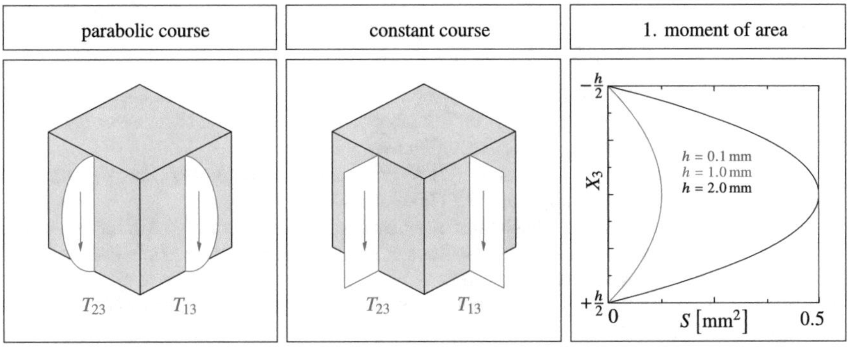

Fig. 3.3 Considerations for the correction of transverse shear stress course for a single layer

thickness, the amplitude thus decreases sharply. A constant comparative measure for this course thus has a better degree of approximation for small structural thicknesses. Thus, for the complementary energies, approximately $W^{\star}_{\text{real}} \approx W^{\star}_{\text{const}}$ holds true and the shear correction factor tends to $\kappa = 1$. The situation is illustrated in Fig. 3.3 on the right-hand side, where various structural thicknesses are compared exemplary.

However, in the composite structure investigated in this work, the transversal shear stress course must also be examined in the overall structure. This is illustrated in Fig. 3.4. Focusing on the core layer, it is noticeable that the variation of the transverse shear stress curve approaches zero. The transverse shear stress in the range $-{}^{h^c}\!/_2 \cdots + {}^{h^c}\!/_2$ is almost constant. The constant distribution introduced in Fig. 3.3 (center) thus reflects the real course in this subrange well. The complementary energies of reality and approximation are approximately equal. For the shear correction factor $\kappa^c \approx 1$ holds true.

The shear correction factor of the two skin layers is relevant due to the relatively high shear moduli G^s and high structural thicknesses h^s of the skin layers and the associated high shear stiffness against the core layer ($D^s_{\text{S}} \gg D^c_{\text{S}}, \forall\, s \in \{t, b\}$). Therefore, $\kappa^K = 1\ \forall\ K \in \{t, c, b\}$ is fully valid for present investigations at Anti-Sandwiches.

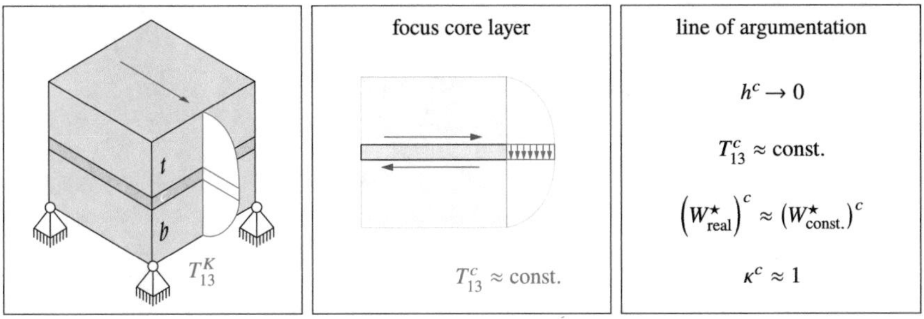

Fig. 3.4 Course of transverse shear stress at an infinitesimal Anti-Sandwich volume element

References

1. Altenbach H, Altenbach J, Rikards R (1996) Einführung in die Mechanik der Laminat- und Sandwichtragwerke - Modellierung und Berechnung von Balken und Platten aus Verbundwerkstoffen. Deutscher Verlag für Grundstoffindustrie, Stuttgart
2. Altenbach H, Altenbach J, Kissing W (2004) Mechanics of composite structural elements. Springer, Berlin. https://doi.org/10.1007/978-3-662-08589-9
3. Aßmus M, Naumenko K, Altenbach H (2016) A multiscale projection approach for the coupled global-local structural analysis of photovoltaic modules. Compos Struct 158:340–358. https://doi.org/10.1016/j.compstruct.2016.09.036
4. Mittelstedt C, Becker W (2017) Strukturmechanik ebener Laminate. Technische Universität Darmstadt, Darmstadt
5. Naumenko K, Eremeyev VA (2014) A layer-wise theory for laminated glass and photovoltaic panels. Compos Struct 112:283–291. https://doi.org/10.1016/j.compstruct.2014.02.009
6. Neff P (2004) A geometrically exact cosserat shell-model including size effects, avoiding degeneracy in the thin shell limit. Part i: formal dimensional reduction for elastic plates and existence of minimizers for positive cosserat couple modulus. Contin Mech Thermodyn 16(6):577–628. https://doi.org/10.1007/s00161-004-0182-4
7. Schulze SH, Pander M, Naumenko K, Altenbach H (2012) Analysis of laminated glass beams for photovoltaic applications. Int J Solids Struct 49(15):2027–2036. https://doi.org/10.1016/j.ijsolstr.2012.03.028
8. Timoshenko SP (1921) On the correction for shear of the differential equation for transverse vibrations of prismatic bars. Philos Mag 41(245):744–746. https://doi.org/10.1080/14786442108636264
9. Timoshenko SP (1922) On the transverse vibrations of bars of uniform cross-section. Philos Mag 43(253):125–131. https://doi.org/10.1080/14786442208633855
10. Weps M, Naumenko K, Altenbach H (2013) Unsymmetric three-layer laminate with soft core for photovoltaic modules. Compos Struct 105:332–339. https://doi.org/10.1016/j.compstruct.2013.05.029

Chapter 4
Vartiational Formulation

4.1 Formal Procedure

In the previous chapters, the local behavior of an elastic, threelayered composite structure was derived, which resulted in the description of the initial boundary value problem. The principle of virtual work is a formulation equivalent to the balances of forces and moments, and represents a weak form of the balance [1]. It is obtained by weighting the equations of motion with test functions equivalent to the vectors of degrees of freedom and subsequent partial integration over the area considered [5]. The test function can be interpreted as infinitesimal deformation field (virtual displacements, virtual deflections, and virtual rotations). This field is arbitrary, but must satisfy the geometric boundary conditions and have to be continuously differentiable [1]. There is no further assumption in this principle [4]. The following process serves as a basis for the numerical implementation.

The equations for force and moment balance which were determined in Chap. 2 serve as a basis.

$$\nabla \cdot \left(\boldsymbol{N}^K + \boldsymbol{q}^K \otimes \boldsymbol{n}\right) \qquad + \boldsymbol{f}^K = \rho^K \left(\ddot{\boldsymbol{a}}^K + (\boldsymbol{J}_{\mathrm{T}}^K)^\top \cdot \boldsymbol{n} \times \ddot{\boldsymbol{\varphi}}^K\right) \tag{4.1}$$

$$\nabla \cdot \left(-\boldsymbol{L}^K \times \boldsymbol{n}\right) + \boldsymbol{q}^K \times \boldsymbol{n} \;\; + \boldsymbol{m}^K = \rho^K \left(\boldsymbol{J}_{\mathrm{T}}^K \cdot \ddot{\boldsymbol{a}}^K + \boldsymbol{J}_{\mathrm{R}}^K \cdot \boldsymbol{n} \times \ddot{\boldsymbol{\varphi}}^K\right) \tag{4.2}$$

In order to gain a clear representation, these equations are again given in analogy to Chap. 3 in form of the equilibria of membrane force, equilibria of transverse shear forces, and the equilibria of moments for every layer separately.

$$\nabla \cdot \boldsymbol{N}^t + \boldsymbol{s}^t - \boldsymbol{s} = \rho^t \left(\ddot{\boldsymbol{v}}^t + \alpha^t \ddot{\boldsymbol{\varphi}}^t\right) \tag{4.3}$$

$$\nabla \cdot \boldsymbol{N}^c + \boldsymbol{s}^b - \boldsymbol{s}^t = \rho^c \ddot{\boldsymbol{v}}^c \tag{4.4}$$

$$\nabla \cdot \boldsymbol{N}^b - \boldsymbol{s}^b = \rho^b \left(\ddot{\boldsymbol{v}}^b + \alpha^b \ddot{\boldsymbol{\varphi}}^b\right) \tag{4.5}$$

$$\nabla \cdot \boldsymbol{q}^t + q^t + q = \rho^t \ddot{w}^t \tag{4.6}$$

$$\nabla \cdot \boldsymbol{q}^c + q^b - q^t = \rho^c \ddot{w}^c \tag{4.7}$$

M. Aßmus, *Structural Mechanics of Anti-Sandwiches*,
SpringerBriefs in Continuum Mechanics,
https://doi.org/10.1007/978-3-030-04354-4_4

$$
\begin{aligned}
\nabla \cdot \boldsymbol{q}^b - q^b &= \rho^b \ddot{w}^b && (4.8)\\
\nabla \cdot \boldsymbol{L}^t - \boldsymbol{q}^t + \frac{h^t}{2}\left(\boldsymbol{s} + \boldsymbol{s}^t\right) &= \rho^t \left(\alpha^t \ddot{\boldsymbol{v}}^t + \beta^t \ddot{\boldsymbol{\varphi}}^t\right) && (4.9)\\
\nabla \cdot \boldsymbol{L}^c - \boldsymbol{q}^c + \frac{h^c}{2}\left(\boldsymbol{s}^t + \boldsymbol{s}^b\right) &= \rho^c \beta^c \ddot{\boldsymbol{\varphi}}^c && (4.10)\\
\nabla \cdot \boldsymbol{L}^b - \boldsymbol{q}^b + \frac{h^b}{2}\boldsymbol{s}^b &= \rho^b \left(\alpha^b \ddot{\boldsymbol{v}}^b + \beta^b \ddot{\boldsymbol{\varphi}}^b\right) && (4.11)
\end{aligned}
$$

At this point, the constraints (3.3), (3.6), and (3.7) are introduced. In doing so, (3.6) and (3.7) are rearranged so that they can be expressed in terms of the degrees of freedom of the core layer.

$$
\boldsymbol{v}^c = \frac{1}{2}\left(\boldsymbol{v}^t + \boldsymbol{v}^b\right) + \frac{1}{4}\left(h^t \boldsymbol{\varphi}^t - h^b \boldsymbol{\varphi}^b\right) \tag{4.12}
$$

$$
\boldsymbol{\varphi}^c = \frac{1}{h^c}\left(\boldsymbol{v}^b - \boldsymbol{v}^t - \frac{h^b}{2}\boldsymbol{\varphi}^b - \frac{h^t}{2}\boldsymbol{\varphi}^t\right) \tag{4.13}
$$

$$
w = w^t = w^c = w^b \tag{4.14}
$$

Now, Eqs. (3.27)–(3.35) are summed up, substituting corresponding expressions by the three above constraints. The virtual degrees of freedom are applied to this construct of equilibrium. This results in the following expression.

$$
\begin{aligned}
&\nabla \cdot \boldsymbol{N}^t \cdot \delta \boldsymbol{v}^t + \nabla \cdot \boldsymbol{N}^b \cdot \delta \boldsymbol{v}^b + \nabla \cdot \boldsymbol{N}^c \cdot \left(\frac{1}{2}\left[\delta \boldsymbol{v}^t + \delta \boldsymbol{v}^b\right] + \frac{1}{4}\left[h^t \delta \boldsymbol{\varphi}^t - h^b \delta \boldsymbol{\varphi}^b\right]\right)\\
&+ \nabla \cdot \left[\boldsymbol{q}^t + \boldsymbol{q}^c + \boldsymbol{q}^b\right] \delta w\\
&+ \nabla \cdot \boldsymbol{L}^t \cdot \delta \boldsymbol{\varphi}^t + \nabla \cdot \boldsymbol{L}^b \cdot \delta \boldsymbol{\varphi}^b + \nabla \cdot \boldsymbol{L}^c \cdot \frac{1}{h^c}\left[\delta \boldsymbol{v}^b - \delta \boldsymbol{v}^t - \frac{h^b}{2}\delta \boldsymbol{\varphi}^b - \frac{h^t}{2}\delta \boldsymbol{\varphi}^t\right]\\
&- \boldsymbol{q}^t \cdot \delta \boldsymbol{\varphi}^t - \boldsymbol{q}^b \cdot \delta \boldsymbol{\varphi}^b - \boldsymbol{q}^c \cdot \frac{1}{h^c}\left[\delta \boldsymbol{v}^b - \delta \boldsymbol{v}^t - \frac{h^b}{2}\delta \boldsymbol{\varphi}^b - \frac{h^t}{2}\delta \boldsymbol{\varphi}^t\right]\\
&+ q\,\delta w - \boldsymbol{s} \cdot \delta \boldsymbol{v}^t + \frac{h^t}{2}\boldsymbol{s} \cdot \delta \boldsymbol{\varphi}^t\\
&= \qquad\qquad\qquad\qquad\qquad\qquad\qquad\qquad\qquad\qquad\qquad\qquad (4.15)\\
&\left[\rho^t + \rho^c + \rho^b\right] \ddot{w}\,\delta w\\
&+ \left\{\rho^t \left[\ddot{\boldsymbol{v}}^t + \alpha^t \ddot{\boldsymbol{\varphi}}^t\right] - \rho^c \beta^c \left(\frac{1}{h^c}\right)^2 \left[\ddot{\boldsymbol{v}}^b - \ddot{\boldsymbol{v}}^t - \frac{h^b}{2}\ddot{\boldsymbol{\varphi}}^b - \frac{h^t}{2}\ddot{\boldsymbol{\varphi}}^t\right]\right.\\
&\left. + \frac{1}{2}\rho^c \left[\frac{1}{2}\left(\ddot{\boldsymbol{v}}^t + \ddot{\boldsymbol{v}}^b\right) + \frac{1}{4}\left(h^t \ddot{\boldsymbol{\varphi}}^t - h^b \ddot{\boldsymbol{\varphi}}^b\right)\right]\right\} \cdot \delta \boldsymbol{v}^t\\
&+ \left\{\rho^b \left[\ddot{\boldsymbol{v}}^b + \alpha^b \ddot{\boldsymbol{\varphi}}^b\right] - \rho^c \beta^c \left(\frac{1}{h^c}\right)^2 \left[\ddot{\boldsymbol{v}}^b - \ddot{\boldsymbol{v}}^t - \frac{h^b}{2}\ddot{\boldsymbol{\varphi}}^b - \frac{h^t}{2}\ddot{\boldsymbol{\varphi}}^t\right]\right.
\end{aligned}
$$

$$
+\frac{1}{2}\rho^c\left[\frac{1}{2}\left(\ddot{v}^t+\ddot{v}^b\right)+\frac{1}{4}\left(h^t\ddot{\varphi}^t-h^b\ddot{\varphi}^b\right)\right]\Bigg\}\cdot\delta v^b
$$

$$
+\left\{\rho^t\left[\alpha^t\ddot{v}^t+\beta^t\ddot{\varphi}^t\right]-\rho^c\beta^c\left(\frac{1}{h^c}\right)^2\frac{h^t}{2}\left[\ddot{v}^b-\ddot{v}^t-\frac{h^b}{2}\ddot{\varphi}^b-\frac{h^t}{2}\ddot{\varphi}^t\right]\right.
$$

$$
\left.+\frac{1}{4}h^t\rho^c\left[\frac{1}{2}\left(\ddot{v}^t+\ddot{v}^b\right)+\frac{1}{4}\left(h^t\ddot{\varphi}^t-h^b\ddot{\varphi}^b\right)\right]\right\}\cdot\delta\varphi^t
$$

$$
+\left\{\rho^b\left[\alpha^b\ddot{v}^b+\beta^b\ddot{\varphi}^b\right]-\rho^c\beta^c\left(\frac{1}{h^c}\right)^2\frac{h^b}{2}\left[\ddot{v}^b-\ddot{v}^t-\frac{h^b}{2}\ddot{\varphi}^b-\frac{h^t}{2}\ddot{\varphi}^t\right]\right.
$$

$$
\left.+\frac{1}{4}h^b\rho^c\left[\frac{1}{2}\left(\ddot{v}^t+\ddot{v}^b\right)+\frac{1}{4}\left(h^t\ddot{\varphi}^t-h^b\ddot{\varphi}^b\right)\right]\right\}\cdot\delta\varphi^b
$$

For a clear description of the following, the abbreviations *LHS* for the left side and *RHS* for the right side are introduced. This simplifies Eq. (4.15) using $LHS = RHS$.

4.2 Introduction of Global Quantities

So far, the quantities $\boldsymbol{v}^K$, $\boldsymbol{\varphi}^K$, ρ^K, α^K, β^K and h^K for all $K \in \{t, c, b\}$ were used. At this point, abbreviations for the degrees of freedom are introduced, which are based on the arithmetic mean (layer index $K = \circ$) or its variation ($K = \Delta$) of the measures of the front and back cover. This introduction is arbitrary and is only intended to serve the compact notation of the following transformations. First, in-plane translations

$$
\boldsymbol{v}^\circ = \frac{1}{2}\left(\boldsymbol{v}^t + \boldsymbol{v}^b\right) \tag{4.16}
$$

$$
\boldsymbol{v}^\Delta = \frac{1}{2}\left(\boldsymbol{v}^t - \boldsymbol{v}^b\right) \tag{4.17}
$$

and out-of-plane rotations

$$
\boldsymbol{\varphi}^\circ = \frac{1}{2}\left(\boldsymbol{\varphi}^t + \boldsymbol{\varphi}^b\right) \tag{4.18}
$$

$$
\boldsymbol{\varphi}^\Delta = \frac{1}{2}\left(\boldsymbol{\varphi}^t - \boldsymbol{\varphi}^b\right) \tag{4.19}
$$

are reformulated. Unique inverse operations exist for these terms

$$
\boldsymbol{v}^t = \boldsymbol{v}^\circ + \boldsymbol{v}^\Delta \qquad \boldsymbol{\varphi}^t = \boldsymbol{\varphi}^\circ + \boldsymbol{\varphi}^\Delta \tag{4.20}
$$

$$
\boldsymbol{v}^b = \boldsymbol{v}^\circ - \boldsymbol{v}^\Delta \qquad \boldsymbol{\varphi}^b = \boldsymbol{\varphi}^\circ - \boldsymbol{\varphi}^\Delta \tag{4.21}
$$

such abbreviations are also introduced for the mass density of the individual layers.

$$\left.\begin{aligned} \rho^\circ &= \rho^t + \rho^b + \rho^c \\ \rho^\Delta &= \rho^t - \rho^b \end{aligned}\right\} \qquad \left\{\begin{aligned} \rho^t &= \frac{1}{2}\left(\rho^\circ + \rho^\Delta - \rho^c\right) \\ \rho^b &= \frac{1}{2}\left(\rho^\circ - \rho^\Delta - \rho^c\right) \end{aligned}\right. \tag{4.22}$$

This also applies to the abbreviations for the translational

$$\left.\begin{aligned} \alpha^\circ &= \frac{1}{2}\left(\alpha^t + \alpha^b\right) \\ \alpha^\Delta &= \frac{1}{2}\left(\alpha^t - \alpha^b\right) \end{aligned}\right\} \qquad \left\{\begin{aligned} \alpha^t &= \alpha^\circ + \alpha^\Delta \\ \alpha^b &= \alpha^\circ - \alpha^\Delta \end{aligned}\right. \tag{4.23}$$

and the rotational terms of inertia.

$$\left.\begin{aligned} \beta^\circ &= \frac{1}{2}\left(\beta^t + \beta^b\right) \\ \beta^\Delta &= \frac{1}{2}\left(\beta^t - \beta^b\right) \end{aligned}\right\} \qquad \left\{\begin{aligned} \beta^t &= \beta^\circ + \beta^\Delta \\ \beta^b &= \beta^\circ - \beta^\Delta \end{aligned}\right. \tag{4.24}$$

The following abbreviations are used for the structural thicknesses

$$\left.\begin{aligned} h^\circ &= \frac{1}{2}\left(h^t + h^b\right) \\ h^\Delta &= \frac{1}{2}\left(h^t - h^b\right) \end{aligned}\right\} \qquad \left\{\begin{aligned} h^t &= h^\circ + h^\Delta \\ h^b &= h^\circ - h^\Delta \end{aligned}\right. , \tag{4.25}$$

whereby the following abbreviation proves also useful to simplify the notation.

$$\tilde{H} = h^c + h^\circ \tag{4.26}$$

In addition, such global variables should also be introduced for the kinematic, kinetic and constitutive measures. The following agreements are made for the stress resultants at composite level.

$$\boldsymbol{N}^\circ = \sum_K \boldsymbol{N}^K \qquad \forall\, K \in \{t, c, b\} \tag{4.27}$$

$$\boldsymbol{L}^\circ = \sum_K \boldsymbol{L}^K + \frac{1}{2}\left[h^c + h^b\right]\boldsymbol{N}^b - \frac{1}{2}\left[h^c + h^t\right]\boldsymbol{N}^t \qquad \forall\, K \in \{t, c, b\} \tag{4.28}$$

$$\boldsymbol{q}^\circ = \sum_K \boldsymbol{q}^K \qquad \forall\, K \in \{t, c, b\} \tag{4.29}$$

$$\boldsymbol{N}^{\Delta} = \boldsymbol{N}^{t} - \boldsymbol{N}^{b} \tag{4.30}$$

$$\boldsymbol{L}^{\Delta} = \boldsymbol{L}^{t} - \boldsymbol{L}^{b} \tag{4.31}$$

$$\boldsymbol{q}^{\Delta} = \boldsymbol{q}^{t} - \boldsymbol{q}^{b} \tag{4.32}$$

These global stress resultants are clearly traceable to the measures of the skin layers.

$$\boldsymbol{N}^{t} = \frac{1}{2}\left[\boldsymbol{N}^{\circ} + \boldsymbol{N}^{\Delta} - \boldsymbol{N}^{c}\right] \qquad \boldsymbol{N}^{b} = \frac{1}{2}\left[\boldsymbol{N}^{\circ} - \boldsymbol{N}^{\Delta} - \boldsymbol{N}^{c}\right] \tag{4.33}$$

$$\boldsymbol{L}^{t} = \frac{1}{2}\left[\boldsymbol{L}^{\circ} + \boldsymbol{L}^{\Delta} - \boldsymbol{L}^{c}\right] + \frac{1}{8}\left(h^{t}-h^{b}\right)\boldsymbol{N}^{\circ} + \frac{1}{8}\left(h^{t}+h^{b}+2h^{c}\right)\boldsymbol{N}^{\Delta} + \frac{1}{8}\left(h^{b}-h^{t}\right)\boldsymbol{N}^{c} \tag{4.34}$$

$$\boldsymbol{L}^{b} = \frac{1}{2}\left[\boldsymbol{L}^{\circ} - \boldsymbol{L}^{\Delta} - \boldsymbol{L}^{c}\right] + \frac{1}{8}\left(h^{t}-h^{b}\right)\boldsymbol{N}^{\circ} + \frac{1}{8}\left(h^{t}+h^{b}+2h^{c}\right)\boldsymbol{N}^{\Delta} + \frac{1}{8}\left(h^{b}-h^{t}\right)\boldsymbol{N}^{c} \tag{4.35}$$

$$\boldsymbol{q}^{t} = \frac{1}{2}\left[\boldsymbol{q}^{\circ} + \boldsymbol{q}^{\Delta} - \boldsymbol{q}^{c}\right] \qquad \boldsymbol{q}^{b} = \frac{1}{2}\left[\boldsymbol{q}^{\circ} - \boldsymbol{q}^{\Delta} - \boldsymbol{q}^{c}\right] \tag{4.36}$$

Here, the already determined measures for the core layer $\boldsymbol{N}^{c}$, $\boldsymbol{L}^{c}$, $\boldsymbol{q}^{c}$ are continued, too. By means of the global variables $K \in \{\circ, \Delta, c\}$, the resultants acting in the direction of the boundary normal can be introduced at the composite level.

$$\boldsymbol{n}_{\nu}^{\circ} = \boldsymbol{\nu} \cdot \boldsymbol{N}^{\circ} \qquad \boldsymbol{n}_{\nu}^{\Delta} = \boldsymbol{\nu} \cdot \boldsymbol{N}^{\Delta} \qquad \boldsymbol{n}_{\nu}^{c} = \boldsymbol{\nu} \cdot \boldsymbol{N}^{c} \tag{4.37}$$

$$\boldsymbol{m}_{\nu}^{\circ} = \boldsymbol{\nu} \cdot \boldsymbol{L}^{\circ} \qquad \boldsymbol{m}_{\nu}^{\Delta} = \boldsymbol{\nu} \cdot \boldsymbol{L}^{\Delta} \qquad \boldsymbol{m}_{\nu}^{c} = \boldsymbol{\nu} \cdot \boldsymbol{L}^{c} \tag{4.38}$$

$$q_{\nu}^{\circ} = \boldsymbol{\nu}\boldsymbol{q}^{\circ} \qquad q_{\nu}^{\Delta} = \boldsymbol{\nu} \cdot \boldsymbol{q}^{\Delta} \qquad q_{\nu}^{c} = \boldsymbol{\nu} \cdot \boldsymbol{q}^{c} \tag{4.39}$$

The resulting surface on which the global structural analysis considerations are based on can be splitted into a loaded $\mathfrak{S}_{\mathrm{N}}^{\circ}$ and a supported boundary $\mathfrak{S}_{\mathrm{D}}^{\circ}$. Corresponding boundary conditions are as follows.

$$^{\star}\boldsymbol{n}_{\nu}^{\circ} = \boldsymbol{n}_{\nu}^{\circ}\big|_{\mathfrak{S}_{\mathrm{N}}^{\circ}} \qquad ^{\star}\boldsymbol{n}_{\nu}^{\Delta} = \boldsymbol{n}_{\nu}^{\Delta}\big|_{\mathfrak{S}_{\mathrm{N}}^{\circ}} \qquad ^{\star}\boldsymbol{n}_{\nu}^{c} = \boldsymbol{n}_{\nu}^{c}\big|_{\mathfrak{S}_{\mathrm{N}}^{\circ}} \tag{4.40}$$

$$^{\star}\boldsymbol{m}_{\nu}^{\circ} = \boldsymbol{m}_{\nu}^{\circ}\big|_{\mathfrak{S}_{\mathrm{N}}^{\circ}} \qquad ^{\star}\boldsymbol{m}_{\nu}^{\Delta} = \boldsymbol{m}_{\nu}^{\Delta}\big|_{\mathfrak{S}_{\mathrm{N}}^{\circ}} \qquad ^{\star}\boldsymbol{m}_{\nu}^{c} = \boldsymbol{m}_{\nu}^{c}\big|_{\mathfrak{S}_{\mathrm{N}}^{\circ}} \tag{4.41}$$

$$^{\star}q_{\nu}^{\circ} = q_{\nu}^{\circ}\big|_{\mathfrak{S}_{\mathrm{N}}^{\circ}} \qquad ^{\star}q_{\nu}^{\Delta} = q_{\nu}^{\Delta}\big|_{\mathfrak{S}_{\mathrm{N}}^{\circ}} \qquad ^{\star}q_{\nu}^{c} = q_{\nu}^{c}\big|_{\mathfrak{S}_{\mathrm{N}}^{\circ}} \tag{4.42}$$

$$^{\star}\boldsymbol{v}^{\circ} = \boldsymbol{v}^{\circ}\big|_{\mathfrak{S}_{\mathrm{D}}^{\circ}} \qquad ^{\star}\boldsymbol{v}^{\Delta} = \boldsymbol{v}^{\Delta}\big|_{\mathfrak{S}_{\mathrm{D}}^{\circ}} \tag{4.43}$$

$$^{\star}\boldsymbol{\varphi}^{\circ} = \boldsymbol{\varphi}^{\circ}\big|_{\mathfrak{S}_{\mathrm{D}}^{\circ}} \qquad ^{\star}\boldsymbol{\varphi}^{\Delta} = \boldsymbol{\varphi}^{\Delta}\big|_{\mathfrak{S}_{\mathrm{D}}^{\circ}} \tag{4.44}$$

$$^{\star}w = w\big|_{\mathfrak{S}_{\mathrm{D}}^{\circ}} \tag{4.45}$$

Variables with a superscript star also represent prescribed quantities at the surface. The variation of the prescribed degrees of freedom is zero.

The constitutive tensors corresponding to (4.27)–(4.32) are introduced as follows.

$$\mathcal{A}^{\circ} = \mathcal{A}^{t} + \mathcal{A}^{b} \tag{4.46}$$

$$\mathcal{A}^{\Delta} = \mathcal{A}^{t} - \mathcal{A}^{b} \tag{4.47}$$

$$\mathcal{D}^{\circ} = \mathcal{D}^{t} + \mathcal{D}^{b} \tag{4.48}$$

$$\mathcal{D}^{\Delta} = \mathcal{D}^{t} - \mathcal{D}^{b} \tag{4.49}$$

$$\boldsymbol{Z}^{\circ} = \boldsymbol{Z}^{t} + \boldsymbol{Z}^{b} \tag{4.50}$$

$$\boldsymbol{Z}^{\Delta} = \boldsymbol{Z}^{t} - \boldsymbol{Z}^{b} \tag{4.51}$$

Also, the stiffness tensors of the core layer $\mathcal{A}^{c}$, $\mathcal{D}^{c}$, $\boldsymbol{Z}^{c}$ are preserved, whereby the indices t, c and b correspond to the layerwise quantities introduced in Chap. 2. The kinematic variables ultimately required are determined as follows.

$$\boldsymbol{G}^{\circ} = \frac{1}{2}\left[\boldsymbol{G}^{t} + \boldsymbol{G}^{b}\right] = \nabla^{\text{sym}}\boldsymbol{v}^{\circ} \tag{4.52}$$

$$\boldsymbol{G}^{\Delta} = \frac{1}{2}\left[\boldsymbol{G}^{t} - \boldsymbol{G}^{b}\right] = \nabla^{\text{sym}}\boldsymbol{v}^{\Delta} \tag{4.53}$$

$$\boldsymbol{K}^{\circ} = \frac{1}{2}\left[\boldsymbol{K}^{t} + \boldsymbol{K}^{b}\right] = \nabla^{\text{sym}}\boldsymbol{\varphi}^{\circ} \tag{4.54}$$

$$\boldsymbol{K}^{\Delta} = \frac{1}{2}\left[\boldsymbol{K}^{t} - \boldsymbol{K}^{b}\right] = \nabla^{\text{sym}}\boldsymbol{\varphi}^{\Delta} \tag{4.55}$$

$$\boldsymbol{g}^{\circ} = \frac{1}{2}\left[\boldsymbol{g}^{t} + \boldsymbol{g}^{b}\right] = \nabla w + \boldsymbol{\varphi}^{\circ} \tag{4.56}$$

$$\boldsymbol{g}^{\Delta} = \frac{1}{2}\left[\boldsymbol{g}^{t} - \boldsymbol{g}^{b}\right] = \boldsymbol{\varphi}^{\Delta} \tag{4.57}$$

For this purpose, the kinematic variables of the core layer $\boldsymbol{G}^{c}$, $\boldsymbol{K}^{c}$, and $\boldsymbol{g}^{c}$ remain in the same way. These kinematics quantities are clearly traceable.

$$\boldsymbol{G}^{t} = \boldsymbol{G}^{\circ} + \boldsymbol{G}^{\Delta} \qquad \boldsymbol{G}^{b} = \boldsymbol{G}^{\circ} - \boldsymbol{G}^{\Delta} \tag{4.58}$$

$$\boldsymbol{K}^{t} = \boldsymbol{K}^{\circ} + \boldsymbol{K}^{\Delta} \qquad \boldsymbol{K}^{b} = \boldsymbol{K}^{\circ} - \boldsymbol{K}^{\Delta} \tag{4.59}$$

$$\boldsymbol{g}^{t} = \boldsymbol{g}^{\circ} + \boldsymbol{g}^{\Delta} \qquad \boldsymbol{g}^{b} = \boldsymbol{g}^{\circ} - \boldsymbol{g}^{\Delta} \tag{4.60}$$

With the aid of the measures formed at global composite level, the constitutive equations can be reformulated.

$$\boldsymbol{N}^{\circ} = \left[\mathcal{A}^{\circ} + \mathcal{A}^{c}\right] : \boldsymbol{G}^{\circ} + \mathcal{A}^{\Delta} : \boldsymbol{G}^{\Delta} + \mathcal{A}^{c} : \left[\frac{1}{2}\left(h^{\Delta}\boldsymbol{K}^{\circ} + h^{\circ}\boldsymbol{K}^{\Delta}\right)\right] \tag{4.61}$$

$$\boldsymbol{N}^{c} = \mathcal{A}^{c} : \left[\boldsymbol{G}^{\circ} + \frac{1}{2}\left(h^{\Delta}\boldsymbol{K}^{\circ} + h^{\circ}\boldsymbol{K}^{\Delta}\right)\right] \tag{4.62}$$

$$N^{\Delta} = \mathcal{A}^{\circ} : E^{\Delta} + \mathcal{A}^{\Delta} : E^{\circ} \tag{4.63}$$

$$\begin{aligned} L^{\circ} = {} & \mathcal{D}^{\circ} : K^{\circ} + \mathcal{D}^{\Delta} : K^{\Delta} - \frac{1}{h^c}\mathcal{D}^{c} : \left[2G^{\Delta} + h^{\circ}K^{\circ} + h^{\Delta}K^{\Delta}\right] \\ & - \frac{1}{2}\mathcal{A}^{\circ} : \left[h^{\Delta}K^{\circ} + \tilde{H}K^{\Delta}\right] - \frac{1}{2}\mathcal{A}^{\Delta} : \left[h^{\Delta}G^{\Delta} + \tilde{H}G^{\circ}\right] \end{aligned} \tag{4.64}$$

$$L^{c} = -\frac{1}{h^c}\mathcal{D}^{c} : \left[2G^{\Delta} + h^{\circ}K^{\circ} + h^{\Delta}K^{\Delta}\right] \tag{4.65}$$

$$L^{\Delta} = \mathcal{D}^{\circ} : K^{\Delta} + \mathcal{D}^{\Delta} : K^{\circ} \tag{4.66}$$

$$q^{\circ} = Z^{\circ} \cdot g^{\circ} + Z^{\Delta} \cdot g^{\Delta} + Z^{c} \cdot \left[\nabla w - \frac{1}{h^c}\left(2v^{\Delta} + h^{\circ}\varphi^{\circ} + h^{\Delta}\varphi^{\Delta}\right)\right] \tag{4.67}$$

$$q^{c} = Z^{c} \cdot \left[\nabla w - \frac{1}{h^c}\left(2v^{\Delta} + h^{\circ}\varphi^{\circ} + h^{\Delta}\varphi^{\Delta}\right)\right] \tag{4.68}$$

$$q^{\Delta} = Z^{\circ} \cdot g^{\Delta} + Z^{\Delta} \cdot g^{\circ} \tag{4.69}$$

In sequence, these abbreviations are applied to Eq. (4.15). For a clear representation, the left- and right-hand side of this equation should be treated separately, whereby calculation steps are identical.

4.3 Left-Hand Side

Within the left-hand side (*LHS*), the expressions with the derivatives from Eq. (4.15) are converted taking into account the product rule, also known as Leibniz rule [2]. Hereinafter the application of the product rule is shown as an example for membrane

$$\nabla \cdot \left[N^K \cdot \delta v^K\right] = \nabla \cdot N^K \cdot \delta v^K + e_i \cdot N^K \cdot \delta v^K_{,i} \tag{4.70}$$

$$\nabla \cdot N^K \cdot \delta v^K = \nabla \cdot \left[N^K \cdot \delta v^K\right] - N^K : \delta E^K \tag{4.71}$$

and transverse shear forces.

$$\nabla \cdot \left[q^K \cdot \delta w\right] = \nabla \cdot q^K \cdot \delta w + e_i \cdot q^K \cdot \delta w_{,i} \tag{4.72}$$

$$\nabla \cdot q^K \cdot \delta w = \nabla \cdot \left[q^K \cdot \delta w\right] - q^K \cdot \delta \nabla w \tag{4.73}$$

Analogous to the membrane terms, identical procedure remains valid for the moments. Taking into account the transformations introduced above, the gradient terms can now be separated.

$$
\begin{aligned}
&\nabla \cdot \Big\{ \boldsymbol{N}^t \cdot \delta \boldsymbol{v}^t + \boldsymbol{N}^b \cdot \delta \boldsymbol{v}^b + \boldsymbol{N}^c \cdot \left(\frac{1}{2} \left[\delta \boldsymbol{v}^t + \delta \boldsymbol{v}^b \right] + \frac{1}{4} \left[h^t \delta \boldsymbol{\varphi}^t - h^b \delta \boldsymbol{\varphi}^b \right] \right) \\
&+ \left[\boldsymbol{q}^t + \boldsymbol{q}^c + \boldsymbol{q}^b \right] \delta w \\
&+ \boldsymbol{L}^t \cdot \delta \boldsymbol{\varphi}^t + \boldsymbol{L}^b \cdot \delta \boldsymbol{\varphi}^b + \boldsymbol{L}^c \cdot \frac{1}{h^c} \left[\delta \boldsymbol{v}^b - \delta \boldsymbol{v}^t - \frac{h^b}{2} \delta \boldsymbol{\varphi}^b - \frac{h^t}{2} \delta \boldsymbol{\varphi}^t \right] \Big\} \\
&- \boldsymbol{q}^t \cdot \delta \boldsymbol{\varphi}^t - \boldsymbol{q}^b \cdot \delta \boldsymbol{\varphi}^b - \boldsymbol{q}^c \cdot \frac{1}{h^c} \left[\delta \boldsymbol{v}^b - \delta \boldsymbol{v}^t - \frac{h^b}{2} \delta \boldsymbol{\varphi}^b - \frac{h^t}{2} \delta \boldsymbol{\varphi}^t \right] \\
&+ q \, \delta w - \boldsymbol{s} \cdot \delta \boldsymbol{v}^t + \frac{h^t}{2} \boldsymbol{s} \cdot \delta \boldsymbol{\varphi}^t \\
&- \boldsymbol{N}^t : \delta \boldsymbol{G}^t + \boldsymbol{N}^b : \delta \boldsymbol{G}^b + \boldsymbol{N}^c : \left(\frac{1}{2} \left[\delta \boldsymbol{G}^t + \delta \boldsymbol{G}^b \right] + \frac{1}{4} \left[h^t \delta \boldsymbol{K}^t - h^b \delta \boldsymbol{K}^b \right] \right) \\
&- \left[\boldsymbol{q}^t + \boldsymbol{q}^c + \boldsymbol{q}^b \right] \cdot \delta \nabla w \\
&- \boldsymbol{L}^t : \delta \boldsymbol{K}^t - \boldsymbol{L}^b : \delta \boldsymbol{K}^b - \boldsymbol{L}^c : \frac{1}{h^c} \left[\delta \boldsymbol{G}^b - \delta \boldsymbol{G}^t - \frac{h^b}{2} \delta \boldsymbol{K}^b - \frac{h^t}{2} \delta \boldsymbol{K}^t \right] \\
&= LHS
\end{aligned} \tag{4.74}
$$

At this point the integral over the surface is determined and the two dimensional Gauss theorem (see p. 111 in Appendix A) is applied to receive the principle of virtual work. For the left-hand side this method leads to the internal and the external virtual work.

$$
\begin{aligned}
\delta W_{\text{int}} = \int_{\mathfrak{S}} \Big\{ & \boldsymbol{N}^t : \delta \boldsymbol{G}^t + \boldsymbol{N}^b : \delta \boldsymbol{G}^b \\
& + \boldsymbol{N}^c : \left(\frac{1}{2} \left[\delta \boldsymbol{G}^t + \delta \boldsymbol{G}^b \right] + \frac{1}{4} \left[h^t \delta \boldsymbol{K}^t - h^b \delta \boldsymbol{K}^b \right] \right) \\
& + \left[\boldsymbol{q}^t + \boldsymbol{q}^c + \boldsymbol{q}^b \right] \cdot \delta \nabla w \\
& + \boldsymbol{L}^t : \delta \boldsymbol{K}^t + \boldsymbol{L}^b : \delta \boldsymbol{K}^b \\
& + \boldsymbol{L}^c : \frac{1}{h^c} \left[\delta \boldsymbol{G}^b - \delta \boldsymbol{G}^t - \frac{h^b}{2} \delta \boldsymbol{K}^b - \frac{h^t}{2} \delta \boldsymbol{K}^t \right] \\
& + \boldsymbol{q}^t \cdot \delta \boldsymbol{\varphi}^t + \boldsymbol{q}^b \cdot \delta \boldsymbol{\varphi}^b \\
& + \boldsymbol{q}^c \cdot \frac{1}{h^c} \left[\delta \boldsymbol{v}^b - \delta \boldsymbol{v}^t - \frac{h^b}{2} \delta \boldsymbol{\varphi}^b - \frac{h^t}{2} \delta \boldsymbol{\varphi}^t \right] \Big\} \, \mathrm{d}\mathfrak{S}
\end{aligned} \tag{4.75}
$$

$$
\begin{aligned}
\delta W_{\text{ext}} = \int_{\partial \mathfrak{S}} \boldsymbol{v} \cdot \Big\{ & \boldsymbol{N}^t \cdot \delta \boldsymbol{v}^t + \boldsymbol{N}^b \cdot \delta \boldsymbol{v}^b \\
& + \boldsymbol{N}^c \cdot \left(\frac{1}{2} \left[\delta \boldsymbol{v}^t + \delta \boldsymbol{v}^b \right] + \frac{1}{4} \left[h^t \delta \boldsymbol{\varphi}^t - h^b \delta \boldsymbol{\varphi}^b \right] \right) \\
& + \left[\boldsymbol{q}^t + \boldsymbol{q}^c + \boldsymbol{q}^b \right] \delta w
\end{aligned}
$$

$$+\boldsymbol{L}^t \cdot \delta\boldsymbol{\varphi}^t + \boldsymbol{L}^b \cdot \delta\boldsymbol{\varphi}^b$$

$$+\boldsymbol{L}^c \cdot \frac{1}{h^c}\left[\delta\boldsymbol{v}^b - \delta\boldsymbol{v}^t - \frac{h^b}{2}\delta\boldsymbol{\varphi}^b - \frac{h^t}{2}\delta\boldsymbol{\varphi}^t\right]\Bigg\}\, \mathrm{d}\partial\mathfrak{S}$$

$$+\int_{\mathfrak{S}}\left\{\frac{h^t}{2}\boldsymbol{s}\cdot\delta\boldsymbol{\varphi}^t + \boldsymbol{s}\cdot\delta\boldsymbol{v}^t + p\delta w\right\}\mathrm{d}\mathfrak{S} \tag{4.76}$$

The virtual internal and the virtual external work can be further simplified by using the global kinematic measures (4.16)–(4.19) and geometry sizes (4.25)–(4.26) introduced.

$$\delta W_{\text{int}} = \int_{\mathfrak{S}}\Bigg\{\boldsymbol{N}^\circ : \left(\delta\boldsymbol{G}^\circ + \frac{1}{2}h^\Delta\delta\boldsymbol{K}^\circ\right) + \boldsymbol{N}^\Delta : \left(\delta\boldsymbol{G}^\Delta + \frac{1}{2}\tilde{H}\delta\boldsymbol{K}^\circ\right) + \boldsymbol{N}^c : \frac{1}{2}h^\circ\delta\boldsymbol{K}^\Delta$$

$$+\boldsymbol{q}^\circ\cdot\delta\boldsymbol{g}^\circ + \boldsymbol{q}^\Delta\cdot\delta\boldsymbol{g}^\Delta - \frac{1}{h^c}\boldsymbol{q}^c\cdot\left(2\delta\boldsymbol{v}^\Delta + \tilde{H}\delta\boldsymbol{\varphi}^\circ + h^\Delta\delta\boldsymbol{\varphi}^\Delta\right)$$

$$+\boldsymbol{L}^\circ : \delta\boldsymbol{K}^\circ + \boldsymbol{L}^\Delta : \delta\boldsymbol{K}^\Delta$$

$$-\frac{1}{h^c}\boldsymbol{L}^c : \left(2\delta\boldsymbol{G}^\Delta + \tilde{H}\delta\boldsymbol{K}^\circ + h^\Delta\delta\boldsymbol{K}^\Delta\right)\Bigg\}\,\mathrm{d}\mathfrak{S} \tag{4.77}$$

$$\delta W_{\text{ext}} = \int_{\partial\mathfrak{S}}\boldsymbol{\nu}\cdot\Bigg\{\boldsymbol{N}^\circ\cdot\left(\delta\boldsymbol{v}^\circ + \frac{1}{2}h^\Delta\delta\boldsymbol{\varphi}^\circ\right) + \boldsymbol{N}^\Delta\cdot\left(\delta\boldsymbol{v}^\Delta + \frac{1}{2}\tilde{H}\delta\boldsymbol{\varphi}^\circ\right) + \boldsymbol{N}^c\cdot\frac{1}{2}h^\circ\delta\boldsymbol{\varphi}^\Delta$$

$$+\boldsymbol{q}^\circ\delta w + \boldsymbol{L}^\circ\cdot\delta\boldsymbol{\varphi}^\circ + \boldsymbol{L}^\Delta\cdot\delta\boldsymbol{\varphi}^\Delta - \boldsymbol{L}^c\cdot\frac{1}{h^c}\left[2\delta\boldsymbol{v}^\Delta + \tilde{H}\delta\boldsymbol{\varphi}^\circ + h^\Delta\delta\boldsymbol{\varphi}^\Delta\right]\Bigg\}\,\mathrm{d}\partial\mathfrak{S}$$

$$+\int_{\mathfrak{S}}\left\{\left(\frac{h^\circ}{2} + \frac{h^\Delta}{2}\right)\boldsymbol{s}\cdot(\delta\boldsymbol{\varphi}^\circ + \delta\boldsymbol{\varphi}^\Delta) + \boldsymbol{s}\cdot(\delta\boldsymbol{v}^\circ + \delta\boldsymbol{v}^\Delta) + p\delta w\right\}\mathrm{d}\mathfrak{S} \tag{4.78}$$

4.4 Right-Hand Side

For the right-hand side, the following expression results when considering global measures for degrees of freedom (4.16)–(4.19), geometry (4.25)–(4.26), mass density (4.22), and terms of inertia (4.23)–(4.24).

$$RHS = \rho^\circ\ddot{w}\,\delta w$$

$$+\Bigg\{\frac{1}{2}\left[\rho^\circ + \rho^\Delta - \rho^c\right]\left[\ddot{\boldsymbol{v}}^\circ + \ddot{\boldsymbol{v}}^\Delta + (\alpha^\circ + \alpha^\Delta)(\ddot{\boldsymbol{\varphi}}^\circ + \ddot{\boldsymbol{\varphi}}^\Delta)\right]$$

$$-\rho^c\beta^c\left(\frac{1}{h^c}\right)^2\left[-2\ddot{\boldsymbol{v}}^\Delta - h^\circ\ddot{\boldsymbol{\varphi}}^\circ - h^\Delta\ddot{\boldsymbol{\varphi}}^\Delta\right]$$

$$+\frac{1}{2}\rho^c\left[\ddot{\boldsymbol{v}}^\circ + \frac{1}{2}\left(h^\Delta\ddot{\boldsymbol{\varphi}}^\circ + h^\circ\ddot{\boldsymbol{\varphi}}^\Delta\right)\right]\Bigg\}\cdot\delta\left(\boldsymbol{v}^\circ + \boldsymbol{v}^\Delta\right)$$

$$+\left\{\frac{1}{2}\left[\rho^{\circ}-\rho^{\Delta}-\rho^{c}\right]\left[\ddot{\boldsymbol{v}}^{\circ}-\ddot{\boldsymbol{v}}^{\Delta}+(\alpha^{\circ}-\alpha^{\Delta})(\ddot{\boldsymbol{\varphi}}^{\circ}-\ddot{\boldsymbol{\varphi}}^{\Delta})\right]\right.$$
$$+\rho^{c}\beta^{c}\left(\frac{1}{h^{c}}\right)^{2}\left[-2\ddot{\boldsymbol{v}}^{\Delta}-h^{\circ}\ddot{\boldsymbol{\varphi}}^{\circ}-h^{\Delta}\ddot{\boldsymbol{\varphi}}^{\Delta}\right]$$
$$\left.+\frac{1}{2}\rho^{c}\left[\ddot{\boldsymbol{v}}^{\circ}+\frac{1}{2}\left(h^{\Delta}\ddot{\boldsymbol{\varphi}}^{\circ}+h^{\circ}\ddot{\boldsymbol{\varphi}}^{\Delta}\right)\right]\right\}\cdot\delta\left(\boldsymbol{v}^{\circ}-\boldsymbol{v}^{\Delta}\right)$$
$$+\left\{\frac{1}{2}\left(\rho^{\circ}+\rho^{\Delta}-\rho^{c}\right)\left[(\alpha^{\circ}+\alpha^{\Delta})(\ddot{\boldsymbol{v}}^{\circ}+\ddot{\boldsymbol{v}}^{\Delta})+(\beta^{\circ}+\beta^{\Delta})(\ddot{\boldsymbol{\varphi}}^{\circ}+\ddot{\boldsymbol{\varphi}}^{\Delta})\right]\right.$$
$$-\frac{1}{2}\rho^{c}\beta^{c}(\frac{1}{h^{c}})^{2}(h^{\circ}+h^{\Delta})\left[-2\ddot{\boldsymbol{v}}^{\Delta}-h^{\circ}\ddot{\varphi}^{\circ}-h^{\Delta}\ddot{\varphi}^{\Delta}\right]$$
$$\left.+\frac{1}{4}(h^{\circ}+h^{\Delta})\rho^{c}\left[\ddot{\boldsymbol{v}}^{\circ}+\frac{1}{2}\left(h^{\Delta}\ddot{\boldsymbol{\varphi}}^{\circ}+h^{\circ}\ddot{\boldsymbol{\varphi}}^{\Delta}\right)\right]\right\}\cdot\delta\left(\boldsymbol{\varphi}^{\circ}+\boldsymbol{\varphi}^{\Delta}\right)$$
$$+\left\{\frac{1}{2}\left(\rho^{\circ}-\rho^{\Delta}-\rho^{c}\right)\left[(\alpha^{\circ}-\alpha^{\Delta})(\ddot{\boldsymbol{v}}^{\circ}-\ddot{\boldsymbol{v}}^{\Delta})+(\beta^{\circ}-\beta^{\Delta})(\ddot{\boldsymbol{\varphi}}^{\circ}-\ddot{\boldsymbol{\varphi}}^{\Delta})\right]\right.$$
$$-\frac{1}{2}\rho^{c}\beta^{c}(\frac{1}{h^{c}})^{2}(h^{\circ}-h^{\Delta})\left[-2\ddot{\boldsymbol{v}}^{\Delta}-h^{\circ}\ddot{\varphi}^{\circ}-h^{\Delta}\ddot{\varphi}^{\Delta}\right]$$
$$\left.-\frac{1}{4}(h^{\circ}-h^{\Delta})\rho^{c}\left[\ddot{\boldsymbol{v}}^{\circ}+\frac{1}{2}\left(h^{\Delta}\ddot{\boldsymbol{\varphi}}^{\circ}+h^{\circ}\ddot{\boldsymbol{\varphi}}^{\Delta}\right)\right]\right\}\cdot\delta\left(\boldsymbol{\varphi}^{\circ}-\boldsymbol{\varphi}^{\Delta}\right)\tag{4.79}$$

At this point the separation of the variations takes place. The above equation is thus as follows.

$$RHS=\rho^{\circ}\ddot{w}\,\delta w$$
$$+\left\{(\rho^{\circ}-\rho^{c})\ddot{\boldsymbol{v}}^{\circ}+\rho^{\Delta}\ddot{\boldsymbol{v}}^{\Delta}+\left[(\rho^{\circ}-\rho^{c})\alpha^{\circ}+\rho^{\Delta}\alpha^{\Delta}\right]\ddot{\boldsymbol{\varphi}}^{\circ}\right.$$
$$+\left[(\rho^{\circ}-\rho^{c})\alpha^{\Delta}+\rho^{\Delta}\alpha^{\circ}\right]\ddot{\boldsymbol{\varphi}}^{\Delta}$$
$$\left.+\rho^{c}\left[\ddot{\boldsymbol{v}}^{\circ}+\frac{1}{2}(h^{\Delta}\ddot{\boldsymbol{\varphi}}^{\circ}+h^{\circ}\ddot{\boldsymbol{\varphi}}^{\Delta})\right]\right\}\cdot\delta\boldsymbol{v}^{\circ}$$
$$+\left\{(\rho^{\circ}-\rho^{c})\ddot{\boldsymbol{v}}^{\Delta}+\rho^{\Delta}\ddot{\boldsymbol{v}}^{\circ}+\left[(\rho^{\circ}-\rho^{c})\alpha^{\Delta}+\rho^{\Delta}\alpha^{\circ}\right]\ddot{\boldsymbol{\varphi}}^{\circ}\right.$$
$$+\left[(\rho^{\circ}-\rho^{c})\alpha^{\circ}+\rho^{\Delta}\alpha^{\Delta}\right]\ddot{\boldsymbol{\varphi}}^{\Delta}$$
$$\left.-2\rho^{c}\beta^{c}\left[-2\ddot{\boldsymbol{v}}^{\Delta}-h^{\Delta}\ddot{\boldsymbol{\varphi}}^{\Delta}-h^{\circ}\ddot{\boldsymbol{\varphi}}^{\circ})\right]\right\}\cdot\delta\boldsymbol{v}^{\Delta}$$
$$+\left\{\left[(\rho^{\circ}-\rho^{c})\alpha^{\circ}+\rho^{\Delta}\alpha^{\Delta}\right]\ddot{\boldsymbol{v}}^{\circ}+\left[(\rho^{\circ}-\rho^{c})\alpha^{\Delta}+\rho^{\Delta}\alpha^{\circ}\right]\ddot{\boldsymbol{v}}^{\Delta}\right.$$
$$+\left[\left(\rho^{\circ}-\rho^{c}\right)\beta^{\circ}+\rho^{\Delta}\beta^{\Delta}\right]\ddot{\boldsymbol{\varphi}}^{\circ}+\left[\left(\rho^{\circ}-\rho^{c}\right)\beta^{\Delta}+\rho^{\Delta}\beta^{\circ}\right]\ddot{\boldsymbol{\varphi}}^{\Delta}$$
$$-\rho^{c}\beta^{c}\left(\frac{1}{h^{c}}\right)^{2}h^{\circ}-\left[-2\ddot{\boldsymbol{v}}^{\Delta}-h^{\circ}\ddot{\boldsymbol{\varphi}}^{\circ}-h^{\Delta}\ddot{\boldsymbol{\varphi}}^{\Delta}\right]$$

$$
\begin{aligned}
&+ \frac{1}{2} h^{\Delta} \rho^{c} \left[\ddot{\boldsymbol{v}}^{\circ} + \frac{1}{2} \left(h^{\Delta} \ddot{\boldsymbol{\varphi}}^{\circ} + h^{\circ} \ddot{\boldsymbol{\varphi}}^{\Delta} \right) \right] \Bigg\} \cdot \delta \boldsymbol{\varphi}^{\circ} \\
+ \Bigg\{ &\left[\left(\rho^{\circ} - \rho^{c} \right) \alpha^{\Delta} + \rho^{\Delta} \alpha^{\circ} \right] \ddot{\boldsymbol{v}}^{\circ} \\
&+ \left[\left(\rho^{\circ} - \rho^{c} \right) \alpha^{\circ} + \rho^{\Delta} \alpha^{\Delta} \right] \ddot{\boldsymbol{v}}^{\Delta} \\
&+ \left[\left(\rho^{\circ} - \rho^{c} \right) \beta^{\Delta} + \rho^{\Delta} \beta^{\circ} \right] \ddot{\boldsymbol{\varphi}}^{\circ} \\
&+ \left[\left(\rho^{\circ} - \rho^{c} \right) \beta^{\circ} + \rho^{\Delta} \beta^{\Delta} \right] \ddot{\boldsymbol{\varphi}}^{\Delta} \\
&- \rho^{c} \beta^{c} \left(\frac{1}{h^{c}} \right)^{2} h^{\Delta} \left[-2 \ddot{\boldsymbol{v}}^{\Delta} - h^{\circ} \ddot{\boldsymbol{\varphi}}^{\circ} - h^{\Delta} \ddot{\boldsymbol{\varphi}}^{\Delta} \right] \\
&+ \frac{1}{2} h^{\circ} \rho^{c} \left[\ddot{\boldsymbol{v}}^{\circ} + \frac{1}{2} \left(h^{\Delta} \ddot{\boldsymbol{\varphi}}^{\circ} + h^{\circ} \ddot{\boldsymbol{\varphi}}^{\Delta} \right) \right] \Bigg\} \cdot \delta \boldsymbol{\varphi}^{\Delta}
\end{aligned}
\tag{4.80}
$$

4.5 Terms of Virtual Work

The virtual internal and virtual external work can now be specified as follows.

$$
\begin{aligned}
\delta W_{\text{int}} = \int_{\mathfrak{S}} \Bigg\{ &\delta \boldsymbol{G}^{\circ} : \boldsymbol{\mathcal{A}}^{\circ} : \boldsymbol{G}^{\circ} + \delta \boldsymbol{G}^{\Delta} : \boldsymbol{\mathcal{A}}^{\circ} : \boldsymbol{G}^{\Delta} + \delta \boldsymbol{G}^{\circ} : \boldsymbol{\mathcal{A}}^{\Delta} : \boldsymbol{G}^{\Delta} \\
&+ \delta \boldsymbol{G}^{\Delta} : \boldsymbol{\mathcal{A}}^{\Delta} : \boldsymbol{G}^{\circ} + \left(\delta \boldsymbol{G}^{\circ} + \frac{1}{2} h^{\Delta} \delta \boldsymbol{K}^{\circ} + \frac{1}{2} h^{\circ} \delta \boldsymbol{K}^{\Delta} \right) : \boldsymbol{\mathcal{A}}^{c} \\
&: \left(\boldsymbol{G}^{\circ} + \frac{1}{2} h^{\Delta} \boldsymbol{K}^{\circ} + \frac{1}{2} h^{\circ} \boldsymbol{K}^{\Delta} \right) + \delta \boldsymbol{g}^{\circ} \cdot \boldsymbol{Z}^{\circ} \cdot \boldsymbol{g}^{\circ} \\
&+ \delta \boldsymbol{g}^{\Delta} \cdot \boldsymbol{Z}^{\circ} \cdot \boldsymbol{g}^{\Delta} + \delta \boldsymbol{g}^{\circ} \cdot \boldsymbol{Z}^{\Delta} \cdot \boldsymbol{g}^{\Delta} + \delta \boldsymbol{g}^{\Delta} \cdot \boldsymbol{Z}^{\circ} \cdot \boldsymbol{g}^{\circ} \\
&+ \left[\delta \boldsymbol{g}^{\circ} - \frac{1}{h^{c}} \left(2 \delta \boldsymbol{v}^{\Delta} + \left(h^{c} + h^{\circ} \right) \delta \boldsymbol{\varphi}^{\circ} + h^{\Delta} \delta \boldsymbol{\varphi}^{\Delta} \right) \right] \cdot \boldsymbol{Z}^{c} \\
&\cdot \left[\boldsymbol{g}^{\circ} - \frac{1}{h^{c}} \left(2 \boldsymbol{v}^{\Delta} + \left(h^{c} + h^{\circ} \right) \boldsymbol{\varphi}^{\circ} + h^{\Delta} \boldsymbol{\varphi}^{\Delta} \right) \right] \\
&+ \delta \boldsymbol{K}^{\circ} : \boldsymbol{\mathcal{D}}^{\circ} : \boldsymbol{K}^{\circ} + \delta \boldsymbol{K}^{\Delta} : \boldsymbol{\mathcal{D}}^{\circ} : \boldsymbol{K}^{\Delta} + \delta \boldsymbol{K}^{\circ} : \boldsymbol{\mathcal{D}}^{\Delta} : \boldsymbol{K}^{\Delta} \\
&+ \delta \boldsymbol{K}^{\Delta} : \boldsymbol{\mathcal{D}}^{\Delta} : \boldsymbol{K}^{\circ} + \frac{1}{\left(h^{c} \right)^{2}} \left(2 \delta \boldsymbol{G}^{\Delta} + h^{\circ} \delta \boldsymbol{K}^{\circ} + h^{\Delta} \delta \boldsymbol{K}^{\Delta} \right) : \boldsymbol{\mathcal{D}}^{c} \\
&: \left(2 \boldsymbol{G}^{\Delta} + h^{\circ} \boldsymbol{K}^{\circ} + h^{\Delta} \boldsymbol{K}^{\Delta} \right) \Bigg\} \mathrm{d}\mathfrak{S}
\end{aligned}
\tag{4.81}
$$

$$
\delta W_{\text{ext}} = \int_{\partial \mathfrak{S}_{\mathrm{p}}} \Bigg\{ \left(\delta \boldsymbol{v}^{\circ} + \frac{1}{2} h^{\Delta} \delta \boldsymbol{\varphi}^{\circ} \right) \cdot \boldsymbol{n}_{\nu}^{\circ} + \left[\delta \boldsymbol{v}^{\Delta} + \frac{1}{2} \left(h^{c} + h^{\circ} \right) \delta \boldsymbol{\varphi}^{\circ} \right] \cdot \boldsymbol{n}_{\nu}^{\Delta}
$$

$$
\begin{aligned}
&+\frac{1}{2}h^{\circ}\delta\boldsymbol{\varphi}^{\Delta}\cdot\boldsymbol{n}^{c}_{\nu}+\delta w q^{\circ}_{\nu}+\delta\boldsymbol{\varphi}^{\circ}\cdot\boldsymbol{m}^{\circ}_{\nu}+\delta\boldsymbol{\varphi}^{\Delta}\cdot\boldsymbol{m}^{\Delta}_{\nu}\\
&-\frac{1}{h^{c}}\left[2\delta\boldsymbol{v}^{\Delta}+\left(h^{c}+h^{\circ}\right)\delta\boldsymbol{\varphi}^{\circ}+h^{\Delta}\delta\boldsymbol{\varphi}^{\Delta}\right]\cdot\boldsymbol{m}^{c}_{\nu}\Big\}\mathrm{d}\,\partial\mathfrak{S}_{\mathrm{p}}\\
&+\int\limits_{\mathfrak{S}_{\mathrm{p}}}\left[\frac{h^{t}}{2}\left(\delta\boldsymbol{\varphi}^{\circ}+\delta\boldsymbol{\varphi}^{\Delta}\right)\cdot\boldsymbol{s}-\left(\delta\boldsymbol{v}^{\circ}+\delta\boldsymbol{v}^{\Delta}\right)\cdot\boldsymbol{s}+\delta w\,p\right]\mathrm{d}\,\mathfrak{S}_{\mathrm{p}}
\end{aligned}
\tag{4.82}
$$

Integrating Eq. (4.80) over the surface results in the virtual dynamic work.

$$
\begin{aligned}
\delta W_{\mathrm{dyn}}=\int\limits_{\mathfrak{S}}&\delta w\,\rho^{\circ}\ddot{w}\\
&+\delta\boldsymbol{v}^{\circ}\cdot\Big\{(\rho^{\circ}-\rho^{c})\ddot{\boldsymbol{v}}^{\circ}+\rho^{\Delta}\ddot{\boldsymbol{v}}^{\Delta}+\left[(\rho^{\circ}-\rho^{c})\alpha^{\circ}+\rho^{\Delta}\alpha^{\Delta}\right]\ddot{\boldsymbol{\varphi}}^{\circ}\\
&\qquad+\left[(\rho^{\circ}-\rho^{c})\alpha^{\Delta}+\rho^{\Delta}\alpha^{\circ}\right]\ddot{\boldsymbol{\varphi}}^{\Delta}\\
&\qquad+\rho^{c}\left[\ddot{\boldsymbol{v}}^{\circ}+\frac{1}{2}(h^{\Delta}\ddot{\boldsymbol{\varphi}}^{\circ}+h^{\circ}\ddot{\boldsymbol{\varphi}}^{\Delta})\right]\Big\}\\
&+\delta\boldsymbol{v}^{\Delta}\cdot\Big\{(\rho^{\circ}-\rho^{c})\ddot{\boldsymbol{v}}^{\Delta}+\rho^{\Delta}\ddot{\boldsymbol{v}}^{\circ}+\left[(\rho^{\circ}-\rho^{c})\alpha^{\Delta}+\rho^{\Delta}\alpha^{\circ}\right]\ddot{\boldsymbol{\varphi}}^{\circ}\\
&\qquad+\left[(\rho^{\circ}-\rho^{c})\alpha^{\circ}+\rho^{\Delta}\alpha^{\Delta}\right]\ddot{\boldsymbol{\varphi}}^{\Delta}\\
&\qquad-2\rho^{c}\beta^{c}\left[-2\ddot{\boldsymbol{v}}^{\Delta}-h^{\Delta}\ddot{\boldsymbol{\varphi}}^{\Delta}-h^{\circ}\ddot{\boldsymbol{\varphi}}^{\circ})\right]\Big\}\\
&+\delta\boldsymbol{\varphi}^{\circ}\cdot\Big\{\left[(\rho^{\circ}-\rho^{c})\alpha^{\circ}+\rho^{\Delta}\alpha^{\Delta}\right]\ddot{\boldsymbol{v}}^{\circ}+\left[(\rho^{\circ}-\rho^{c})\alpha^{\Delta}+\rho^{\Delta}\alpha^{\circ}\right]\ddot{\boldsymbol{v}}^{\Delta}\\
&\qquad+\left[\left(\rho^{\circ}-\rho^{c}\right)\beta^{\circ}+\rho^{\Delta}\beta^{\Delta}\right]\ddot{\boldsymbol{\varphi}}^{\circ}+\left[\left(\rho^{\circ}-\rho^{c}\right)\beta^{\Delta}+\rho^{\Delta}\beta^{\circ}\right]\ddot{\boldsymbol{\varphi}}^{\Delta}\\
&\qquad-\rho^{c}\beta^{c}\left(\frac{1}{h^{c}}\right)^{2}h^{\circ}-\left[-2\ddot{\boldsymbol{v}}^{\Delta}-h^{\circ}\ddot{\boldsymbol{\varphi}}^{\circ}-h^{\Delta}\ddot{\boldsymbol{\varphi}}^{\Delta}\right]\\
&\qquad+\frac{1}{2}h^{\Delta}\rho^{c}\left[\ddot{\boldsymbol{v}}^{\circ}+\frac{1}{2}\left(h^{\Delta}\ddot{\boldsymbol{\varphi}}^{\circ}+h^{\circ}\ddot{\boldsymbol{\varphi}}^{\Delta}\right)\right]\Big\}\\
&+\delta\boldsymbol{\varphi}^{\Delta}\cdot\Big\{\left[\left(\rho^{\circ}-\rho^{c}\right)\alpha^{\Delta}+\rho^{\Delta}\alpha^{\circ}\right]\ddot{\boldsymbol{v}}^{\circ}\\
&\qquad+\left[\left(\rho^{\circ}-\rho^{c}\right)\alpha^{\circ}+\rho^{\Delta}\alpha^{\Delta}\right]\ddot{\boldsymbol{v}}^{\Delta}\\
&\qquad+\left[\left(\rho^{\circ}-\rho^{c}\right)\beta^{\Delta}+\rho^{\Delta}\beta^{\circ}\right]\ddot{\boldsymbol{\varphi}}^{\circ}\\
&\qquad+\left[\left(\rho^{\circ}-\rho^{c}\right)\beta^{\circ}+\rho^{\Delta}\beta^{\Delta}\right]\ddot{\boldsymbol{\varphi}}^{\Delta}\\
&\qquad-\rho^{c}\beta^{c}\left(\frac{1}{h^{c}}\right)^{2}h^{\Delta}\left[-2\ddot{\boldsymbol{v}}^{\Delta}-h^{\circ}\ddot{\boldsymbol{\varphi}}^{\circ}-h^{\Delta}\ddot{\boldsymbol{\varphi}}^{\Delta}\right]
\end{aligned}
$$

$$+\frac{1}{2}h^{\circ}\rho^{c}\left[\ddot{\boldsymbol{v}}^{\circ}+\frac{1}{2}\left(h^{\Delta}\ddot{\boldsymbol{\varphi}}^{\circ}+h^{\circ}\ddot{\boldsymbol{\varphi}}^{\Delta}\right)\right]\Bigg\}\,\mathrm{d}\mathfrak{S} \tag{4.83}$$

The three preceding equations define the virtual work terms for the finite element. This allows the entire virtual work to be determined. There are nine kinematic degrees of freedom. These are the average in-plane displacements $v_1^{\circ}\,v_2^{\circ}$, the relative in-plane displacements $v_1^{\Delta}\,v_2^{\Delta}$, the average cross-section rotations $\varphi_1^{\circ}\,\varphi_2^{\circ}$, the relative cross-section rotations $\varphi_1^{\Delta}\,\varphi_2^{\Delta}$ and the deflection w. Since all layers have been modeled on the basis of a five parameter theory, analogous to Mindlins approach [3], the potential energy depends only on the degrees of freedom themselves and their derivatives [5]. Consequently, this model also leads to a C^0 continuous variation problem. In the subsequent chapter, the terms of virtual work derived here are used to derive components of a discretized equation of motion, wherein the transfer takes place on the basis of an orthonormal coordinate system.

References

1. Bathe KJ (2002) Finite-elemente-methoden, 2nd edn. Springer, Berlin
2. Leibniz GW (2005) The early mathematical manuscripts of leibniz. Dover Publications, New York. Translated and with an Introduction by JM Child
3. Mindlin RD (1951) Influence of rotatory inertia and shear on flexural motions of isotropic, elastic plates. J Appl Mech 18:31–38
4. Oden JT, Reddy JN (1983) Variational methods in theoretical mechanics, 2nd edn. Springer, Berlin. https://doi.org/10.1007/978-3-642-68811-9
5. Oñate E (2013) Structural analysis with the finite element method linear statics: vol 2. Beams, plates and shells. Springer, Dordrecht. https://doi.org/10.1007/978-1-4020-8743-1_6

Chapter 5
Finite Element Implementation

5.1 Finite Element Discretization

The weak formulation of the boundary value problem serves as a starting point for the numerical solution of the problem. In this chapter, a spatial finite element discretization is used to generate a semidiscrete structural equation. Choosing a formulation considering the transverse shear distortions is also chosen due to the lower and easier-to-fulfill continuity requirements of the shear flexible finite elements (C^0 continuity) compared to the requirement for shear-rigid elements (C^1 continuity) [6]. Thus only the zeroth derivative of the degrees of freedom has to be continuous. In contrast to the classical, shear-soft Bathe–Dvorkin element [2], in-plane displacements are taken into account. The implementation is based on the global degrees of freedom introduced in the previous chapter.

The finite element method is based on the separation of structure and element level. The structure under investigation Ω is divided in individual domains, i.e. the finite elements Ω^e.

$$\Omega = \bigcup_{e=1}^{NE} \Omega^e \qquad \Omega^i \cap \Omega^j = \emptyset \quad \text{for } i \neq j \tag{5.1}$$

Herein NE indicates the number of elements in the domain Ω. This discretization is illustrated in Fig. 5.1. The boundary of the domain is designated by $\Gamma = \partial\Omega$, whereby the Dirichlet boundary is declared with Γ_{D} and the Neumann boundary with Γ_{N}. The following relationships apply.

$$\Gamma = \Gamma_{\mathrm{D}} \cup \Gamma_{\mathrm{N}} \qquad \Gamma_{\mathrm{D}} \cap \Gamma_{\mathrm{N}} = \emptyset \tag{5.2}$$

These disjoint subsets of the domains boundary are used to specify boundary conditions in the form of primary or dependent variables. In the context of the present work, these are in-plane displacements, rotations and deflections at Γ_{D} as well as boundary resultants of membrane forces, moments and transverse shear forces at Γ_{N}.

M. Aßmus, *Structural Mechanics of Anti-Sandwiches*,
SpringerBriefs in Continuum Mechanics,
https://doi.org/10.1007/978-3-030-04354-4_5

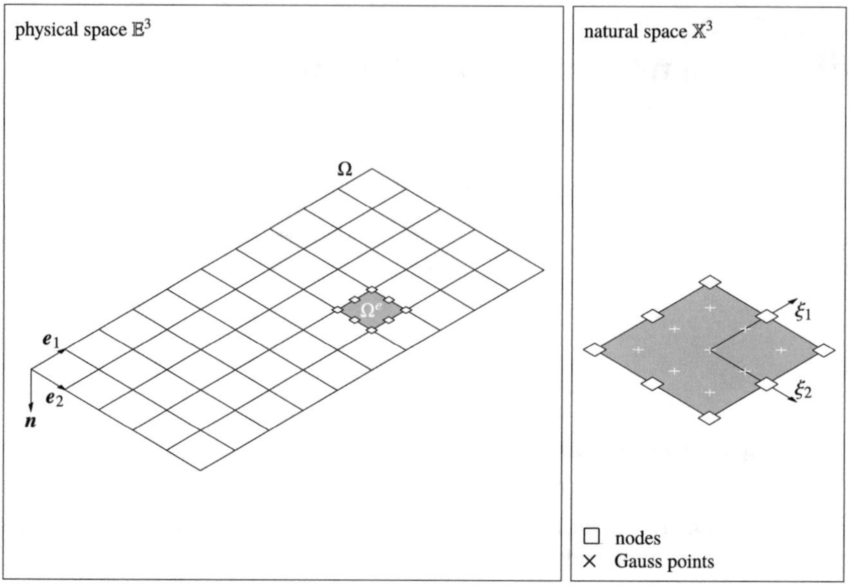

Fig. 5.1 Discretized domain Ω and isoparametric mapping of a finite element Ω^e

In such a discretized structure, the principle of virtual work must be fulfilled in the entire domain as well as in each individual sub-domain.

$$\delta W_{\text{int}} + \delta W_{\text{dyn}} = \delta W_{\text{ext}} \qquad \delta W^e_{\text{int}} + \delta W^e_{\text{dyn}} = \delta W^e_{\text{ext}} \tag{5.3}$$

The virtual works of the overall domain are formed by adding up the individual contributions of the sub-domains.

$$\delta W_{\text{int}} = \sum_{e=1}^{NE} \delta W^e_{\text{int}} \qquad \delta W_{\text{ext}} = \sum_{e=1}^{NE} \delta W^e_{\text{ext}} \qquad \delta W_{\text{dyn}} = \sum_{e=1}^{NE} \delta W^e_{\text{dyn}} \tag{5.4}$$

5.2 Approximation of Field Quantities

The geometric description is here made by the position vector of a material point of a two dimensional surface.

$$\mathbf{x} = \left[X_1 \; X_2\right]^\top \tag{5.5}$$

The introduction of the position vector, in such way defined for only one reference surface is possible since the degrees of freedom introduced in Chap. 4 refer to the middle surface of the composite according to the global coordinate system defined in Chap. 3. Thus, all layers can be represented by only one finite element in the thickness direction. To discretize the equations, the vector of degrees of freedom for each node i of a finite element is initially specified. By introducing the kinematic constraints in Sect. 3.4 as well as the transformation into global quantities in Sect. 4.2, it can be built as follows.

$$\mathbf{a}^i = \left[v_1^{\circ^i} \; v_2^{\circ^i} \; v_1^{\Delta^i} \; v_2^{\Delta^i} \; w^i \; \varphi_1^{\circ^i} \; \varphi_2^{\circ^i} \; \varphi_1^{\Delta^i} \; \varphi_2^{\Delta^i} \right]^\top \qquad \forall\, i = \left\{1, \ldots, N\!N(\Omega^e)\right\} \tag{5.6}$$

The order of degrees of freedom introduced here is arbitrary, but binding for the following embodiments. Here $N\!N(\Omega^e)$ is the number of nodes per element. All node vectors of the degrees of freedom $\mathbf{a}^i$ are combined in the element vector of the degrees of freedom $\mathbf{a}^e$ whereby the vectors per element are sorted by node by node one below the other.

$$\mathbf{a}^e = \left[\mathbf{a}^1 \; \mathbf{a}^2 \; \mathbf{a}^3 \ldots \mathbf{a}^{N\!N(\Omega^e)} \right]^\top \tag{5.7}$$

In order to obtain the degrees of freedom $\mathbf{a}$ as field quantities in element coordinates $\boldsymbol{\xi}$, the element degrees of freedom are identified by means of suitable interpolation functions $\mathbf{N}$, which are characterized by the shape of the finite elements and the approximation order. This interpolation is performed in the coordinates of the finite element $-1 \leq \xi_\alpha \leq 1 \; \forall\, \alpha \in \{1, 2\}$, where the vector of the natural coordinates $\boldsymbol{\xi} = [\,\xi_1 \;\, \xi_2\,]^\top$ is introduced [7, 8].

$$\begin{aligned} \mathbf{a}(\boldsymbol{\xi}) &= \left[v_1^{\circ}(\boldsymbol{\xi}) \; v_2^{\circ}(\boldsymbol{\xi}) \; v_1^{\Delta}(\boldsymbol{\xi}) \; v_2^{\Delta}(\boldsymbol{\xi}) \; w(\boldsymbol{\xi}) \; \varphi_1^{\circ}(\boldsymbol{\xi}) \; \varphi_2^{\circ}(\boldsymbol{\xi}) \; \varphi_1^{\Delta}(\boldsymbol{\xi}) \; \varphi_2^{\Delta}(\boldsymbol{\xi}) \right]^\top \\ &\approx \mathbf{N}(\boldsymbol{\xi})\, \mathbf{a}^e \end{aligned} \tag{5.8}$$

Herein, $\mathbf{N}$ is the matrix that contains the interpolation or shape functions per node.

$$\mathbf{N}(\boldsymbol{\xi}) = \left[\mathbf{N}^1(\boldsymbol{\xi}) \; \mathbf{N}^2(\boldsymbol{\xi}) \; \mathbf{N}^3(\boldsymbol{\xi}) \ldots \mathbf{N}^{N\!N(\Omega^e)}(\boldsymbol{\xi}) \right] \quad \forall\, i \in \left\{1, \ldots, N\!N(\Omega^e)\right\} \tag{5.9}$$

The matrices of the shape functions for the respective node can be formed by means of the shape function function to be selected.

$$\mathbf{N}^i(\boldsymbol{\xi}) = N^i(\boldsymbol{\xi})\, \mathbf{I} \tag{5.10}$$

Here, $\mathbf{I}$ is a quadratic unit matrix, where the number of columns and rows is the same as the number of degrees of freedom per node. This description is also known as isoparametric element concept.

Basically, it remains to be noted that the choice of $\mathbf{a}$ denoting the numerical vector of degrees of freedom is independent of the tensorial designation of the translatoric degrees of freedom of a material point $\boldsymbol{a}$ of the surface continuum as done in Chap. 2.

5.3 Shape Functions

Since the exact values of the degrees of freedom in the total Ω and elementary domain Ω^e are unknown, we introduce shape functions N^i as shown in Eq. (5.8) to approximate the solution. The finite element introduced here uses Serendipity-type shape functions. For this purpose, we introduce some functions of the polynomial degree $PG = 2$ [7].

$$N^i(\boldsymbol{\xi}) = \frac{1}{4}\left(1+\xi_1^i\xi_1\right)\left(1+\xi_2^i\xi_2\right)\left(\xi_1^i\xi_1+\xi_2^i\xi_2-1\right) \qquad i \in \{1,\ldots,4\} \quad (5.11)$$

$$N^i(\boldsymbol{\xi}) = \frac{1}{2}\left(1+\xi_1^i\xi_1\right)\left(1- \quad \xi_2^2\right) \qquad i \in \{6,8\} \quad (5.12)$$

$$N^i(\boldsymbol{\xi}) = \frac{1}{2}\left(1+\xi_2^i\xi_2\right)\left(1- \quad \xi_1^2\right) \qquad i \in \{5,7\} \quad (5.13)$$

The index i here represents the node number of the finite element. The node numbering convention is shown in Fig. 5.2. In addition, the introduced shape functions are visualized there. It can be seen that the element with 8 nodes has a total of 72 degrees of freedom.

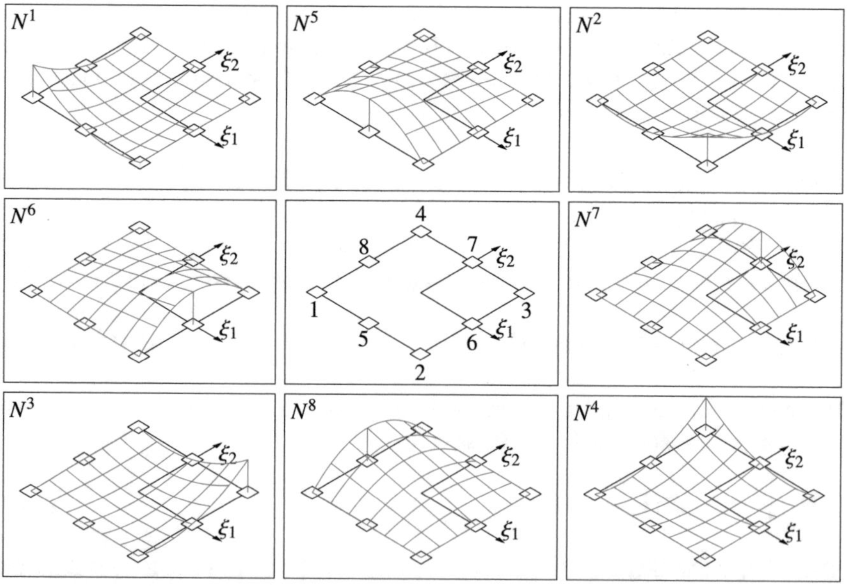

Fig. 5.2 Quadratic shape functions of Serendipity-Type at the plane finite element

5.4 Jacobi Transformation

The transformation of differential line elements $\mathrm{d}X_\alpha$ and $\mathrm{d}\xi_\alpha$ is based on the derivations of the physical and natural coordinates. The transformation is done using the Jacobi matrix $\mathbf{J}(\boldsymbol{\xi})$ and their inverse $\mathbf{J}(\boldsymbol{\xi})^{-1}$ performed. In two dimensions, this operation can be represented as follows [6].

$$\frac{\partial}{\partial \boldsymbol{\xi}} = \mathbf{J}(\boldsymbol{\xi})\,\frac{\partial}{\partial \mathbf{x}} \qquad \frac{\partial}{\partial \mathbf{x}} = \mathbf{J}(\boldsymbol{\xi})^{-1}\,\frac{\partial}{\partial \boldsymbol{\xi}} \tag{5.14}$$

The Jacobi matrices and the individual derivatives are as follows.

$$\mathbf{J}(\boldsymbol{\xi}) = \begin{bmatrix} \dfrac{\partial X_1}{\partial \xi_1} & \dfrac{\partial X_2}{\partial \xi_1} \\ \dfrac{\partial X_1}{\partial \xi_2} & \dfrac{\partial X_2}{\partial \xi_2} \end{bmatrix} \qquad \mathbf{J}(\boldsymbol{\xi})^{-1} = \frac{1}{|\mathbf{J}(\boldsymbol{\xi})|}\begin{bmatrix} \dfrac{\partial X_2}{\partial \xi_2} & -\dfrac{\partial X_2}{\partial \xi_1} \\ -\dfrac{\partial X_1}{\partial \xi_2} & \dfrac{\partial X_1}{\partial \xi_1} \end{bmatrix} \tag{5.15}$$

$$\frac{\partial}{\partial \boldsymbol{\xi}} = \begin{bmatrix} \dfrac{\partial}{\partial \xi_1} & \dfrac{\partial}{\partial \xi_2} \end{bmatrix}^{\mathsf{T}} \qquad \frac{\partial}{\partial \mathbf{x}} = \begin{bmatrix} \dfrac{\partial}{\partial X_1} & \dfrac{\partial}{\partial X_2} \end{bmatrix}^{\mathsf{T}} \tag{5.16}$$

Furthermore, the determinant of the Jacobi matrix $|\mathbf{J}(\boldsymbol{\xi})|$ serves to transform an infinitesimal surface element $\mathrm{d}\Omega$ into physical coordinates to an infinitesimal surface element in natural coordinates.

$$\mathrm{d}\Omega = \mathrm{d}X_1\,\mathrm{d}X_2 = |\mathbf{J}(\boldsymbol{\xi})|\;\mathrm{d}\xi_1\,\mathrm{d}\xi_2 \tag{5.17}$$

5.5 Kinematical Relations

At this point, the kinematic Eqs. (4.52)–(4.57) are discretized. For this purpose, kinematic vectors are introduced with respect to element measures. The quantities responsible for transverse shear are treated separately.

$$\mathbf{e}_{\mathrm{MB}}(\boldsymbol{\xi}) = \left[\mathbf{e}^{\circ}_{\mathrm{M}}\;\mathbf{e}^{\vartriangle}_{\mathrm{M}}\;\mathbf{e}^{\circ}_{\mathrm{B}}\;\mathbf{e}^{\vartriangle}_{\mathrm{B}}\right]^{\mathsf{T}} \qquad \mathbf{e}_{\mathrm{S}}(\boldsymbol{\xi}) = \left[\mathbf{e}^{\circ}_{\mathrm{S}}\;\mathbf{e}^{\vartriangle}_{\mathrm{S}}\right]^{\mathsf{T}} \tag{5.18}$$

It contains the following auxiliary vectors for membrane and transverse shear stains distortions as well as curvature changes.

$$\mathbf{e}^{\circ}_{\mathrm{M}} = \left[G^{\circ}_{11}\;G^{\circ}_{22}\;2G^{\circ}_{12}\right]^{\mathsf{T}} = \left[v^{\circ}_{1,1}\;v^{\circ}_{2,2}\;v^{\circ}_{1,2} + v^{\circ}_{2,1}\right]^{\mathsf{T}} \tag{5.19}$$

$$\mathbf{e}^{\vartriangle}_{\mathrm{M}} = \left[G^{\vartriangle}_{11}\;G^{\vartriangle}_{22}\;2G^{\vartriangle}_{12}\right]^{\mathsf{T}} = \left[v^{\vartriangle}_{1,1}\;v^{\vartriangle}_{2,2}\;v^{\vartriangle}_{1,2} + v^{\vartriangle}_{2,1}\right]^{\mathsf{T}} \tag{5.20}$$

$$\mathbf{e}_{\mathrm{B}}^{\circ} = \left[K_{11}^{\circ}\ K_{22}^{\circ}\ 2K_{12}^{\circ}\right]^{\top} = \left[\varphi_{1,1}^{\circ}\ \varphi_{2,2}^{\circ}\ \varphi_{1,2}^{\circ} + \varphi_{2,1}^{\circ}\right]^{\top} \tag{5.21}$$

$$\mathbf{e}_{\mathrm{B}}^{\vartriangle} = \left[K_{11}^{\vartriangle}\ K_{22}^{\vartriangle}\ 2K_{12}^{\vartriangle}\right]^{\top} = \left[\varphi_{1,1}^{\vartriangle}\ \varphi_{2,2}^{\vartriangle}\ \varphi_{1,2}^{\vartriangle} + \varphi_{2,1}^{\vartriangle}\right]^{\top} \tag{5.22}$$

$$\mathbf{e}_{\mathrm{S}}^{\circ} = \left[g_{1}^{\circ}\ g_{2}^{\circ}\right]^{\top} = \left[w_{,1} + \varphi_{1}^{\circ}\ w_{,2} + \varphi_{2}^{\circ}\right]^{\top} \tag{5.23}$$

$$\mathbf{e}_{\mathrm{S}}^{\vartriangle} = \left[g_{1}^{\vartriangle}\ g_{2}^{\vartriangle}\right]^{\top} = \left[\varphi_{1}^{\vartriangle}\ \varphi_{2}^{\vartriangle}\right]^{\top} \tag{5.24}$$

All the above measures represent fields. The vectors therefore depend on the natural coordinates. For reasons of space and brevity, these dependencies were not explicitly stated. These kinematic fields are approximated in analogy to Eq. (5.8). For this purpose, $\mathbf{B}$ matrices are introduced for the specific deformation parts.

$$\mathbf{e}_{\mathrm{MB}}(\boldsymbol{\xi}) \approx \mathbf{B}_{\mathrm{MB}}(\boldsymbol{\xi})\,\mathbf{a}^{e} \qquad \mathbf{e}_{\mathrm{S}}(\boldsymbol{\xi}) \approx \mathbf{B}_{\mathrm{S}}(\boldsymbol{\xi})\,\mathbf{a}^{e} \tag{5.25}$$

These $\mathbf{B}$ matrices are products of the differential operator $\mathbf{D}$ and the matrix of the shape functions $\mathbf{N}$.

$$\mathbf{B}_{\mathrm{MB}}(\boldsymbol{\xi}) = \mathbf{D}_{\mathrm{MB}}\mathbf{N}(\boldsymbol{\xi}) \qquad \mathbf{B}_{\mathrm{S}}(\boldsymbol{\xi}) = \mathbf{D}_{\mathrm{S}}\mathbf{N}(\boldsymbol{\xi}) \tag{5.26}$$

The structure of the $\mathbf{B}$ and $\mathbf{D}$ matrices is given in Appendix C.

5.6 Constitutive Equations

To implement the constitutive laws, the expressions of the virtual work derived in Chap. 4 must be converted into vector-matrix notation. The basic procedure for transferring kinetic and kinematic measures as well as constitutive quantities is shown in the Appendix C. The constitutive equations of the composite level from Eqs. (4.61)–(4.69) ($\circ$, Δ) and Eqs. (2.43)–(2.45) (c) are as follows.

$$\mathbf{s}_{\mathrm{M}}^{\circ} = \left(\hat{\mathbf{C}}_{\mathrm{M}}^{\circ} + \hat{\mathbf{C}}_{\mathrm{M}}^{c}\right)\mathbf{e}_{\mathrm{M}}^{\circ} + \hat{\mathbf{C}}_{\mathrm{M}}^{\vartriangle}\mathbf{e}_{\mathrm{M}}^{\vartriangle} + \frac{1}{2}\hat{\mathbf{C}}_{\mathrm{M}}^{c}\left(h^{\vartriangle}\mathbf{e}_{\mathrm{B}}^{\circ} + h^{\circ}\mathbf{e}_{\mathrm{B}}^{\vartriangle}\right) \tag{5.27}$$

$$\mathbf{s}_{\mathrm{M}}^{\vartriangle} = \hat{\mathbf{C}}_{\mathrm{M}}^{\circ}\mathbf{e}_{\mathrm{M}}^{\vartriangle} + \hat{\mathbf{C}}_{\mathrm{M}}^{\vartriangle}\mathbf{e}_{\mathrm{M}}^{\circ} \tag{5.28}$$

$$\mathbf{s}_{\mathrm{M}}^{c} = \hat{\mathbf{C}}_{\mathrm{M}}^{c}\left[\mathbf{e}_{\mathrm{M}}^{\circ} + \frac{1}{2}\left(h^{\vartriangle}\mathbf{e}_{\mathrm{B}}^{\circ} + h^{\circ}\mathbf{e}_{\mathrm{B}}^{\vartriangle}\right)\right] \tag{5.29}$$

$$\mathbf{s}_{\mathrm{S}}^{\circ} = \hat{\mathbf{C}}_{\mathrm{S}}^{\circ}\mathbf{e}_{\mathrm{S}}^{\circ} + \hat{\mathbf{C}}_{\mathrm{S}}^{\vartriangle}\mathbf{e}_{\mathrm{S}}^{\vartriangle} + \hat{\mathbf{C}}_{\mathrm{S}}^{c}\left(\mathbf{e}_{\mathrm{S}}^{\circ} + \mathbf{A}_{\mathrm{I}}\mathbf{a}\right) \tag{5.30}$$

$$\mathbf{s}_{\mathrm{S}}^{\vartriangle} = \hat{\mathbf{C}}_{\mathrm{S}}^{\circ}\mathbf{e}_{\mathrm{S}}^{\vartriangle} + \hat{\mathbf{C}}_{\mathrm{S}}^{\vartriangle}\mathbf{e}_{\mathrm{S}}^{\circ} \tag{5.31}$$

$$\mathbf{s}_{\mathrm{S}}^{c} = \hat{\mathbf{C}}_{\mathrm{S}}^{c}\left(\mathbf{e}_{\mathrm{S}}^{\circ} + \mathbf{A}_{\mathrm{I}}\mathbf{a}\right) \tag{5.32}$$

$$\mathbf{s}_{\mathrm{B}}^{\circ} = \hat{\mathbf{C}}_{\mathrm{B}}^{\circ}\mathbf{e}_{\mathrm{B}}^{\circ} + \hat{\mathbf{C}}_{\mathrm{B}}^{\Delta}\mathbf{e}_{\mathrm{B}}^{\Delta} - \frac{1}{h^{c}}\hat{\mathbf{C}}_{\mathrm{B}}^{c}\left(2\mathbf{e}_{\mathrm{M}}^{\Delta} + h^{\circ}\mathbf{e}_{\mathrm{B}}^{\circ} + h^{\Delta}\mathbf{e}_{\mathrm{B}}^{\Delta}\right)$$
$$- \frac{1}{2}\hat{\mathbf{C}}_{\mathrm{M}}^{\circ}\left[h^{\Delta}\mathbf{e}_{\mathrm{M}}^{\circ} + \left(h^{c} + h^{\circ}\right)\mathbf{e}_{\mathrm{M}}^{\Delta}\right] - \frac{1}{2}\hat{\mathbf{C}}_{\mathrm{M}}^{\Delta}\left[h^{\Delta}\mathbf{e}_{\mathrm{M}}^{\Delta} + \left(h^{c} + h^{\circ}\right)\mathbf{e}_{\mathrm{M}}^{\circ}\right] \tag{5.33}$$

$$\mathbf{s}_{\mathrm{B}}^{\Delta} = \hat{\mathbf{C}}_{\mathrm{B}}^{\circ}\mathbf{e}_{\mathrm{B}}^{\Delta} + \hat{\mathbf{C}}_{\mathrm{B}}^{\Delta}\mathbf{e}_{\mathrm{B}}^{\circ} \tag{5.34}$$

$$\mathbf{s}_{\mathrm{B}}^{c} = -\frac{1}{h^{c}}\hat{\mathbf{C}}_{\mathrm{B}}^{c}\left(2\mathbf{e}_{\mathrm{M}}^{\Delta} + h^{\circ}\mathbf{e}_{\mathrm{B}}^{\circ} + h^{\Delta}\mathbf{e}_{\mathrm{B}}^{\Delta}\right) \tag{5.35}$$

The constitutive matrices $\hat{\mathbf{C}}_{\mathrm{M}}^{K}$, $\hat{\mathbf{C}}_{\mathrm{B}}^{K}$, $\hat{\mathbf{C}}_{\mathrm{S}}^{K}$ $\forall K \in \{\circ, \Delta, c\}$ and the auxiliary matrix $\mathbf{A}_1$ introduced here are enclosed in Appendix C.

5.7 Spatial Discretization of Virtual Work

The terms of virtual work (4.81), (4.82), and (4.83) are compiled considering the introduced kinematic and kinetic resultants into the vector-matrix notation, while considering Eqs. (5.8) and (5.25). The integrals are executed with respect to the element surface Ω^e or the element boundary $\partial\Omega^e$. Since this discrete form only applies to the element level, the expressions are indexed with a superscript e. The virtual internal work from Eq. (4.81) can be rewritten as follows.

$$\begin{aligned}\delta W_{\mathrm{int}}^{e} = \int\limits_{\Omega^e} \delta\mathbf{a}^{e\top}\Big[&\mathbf{B}_{\mathrm{S}}^{\top}\Big(\mathbf{C}_{\mathrm{S}}^{\circ} + \mathbf{C}_{\mathrm{S}}^{\Delta} + \mathbf{C}_{\mathrm{S}}^{\Delta\top} + \mathbf{A}_2^{\top}\hat{\mathbf{C}}_{\mathrm{S}}^{c}\mathbf{A}_2\Big)\mathbf{B}_{\mathrm{S}} \\ &+ \mathbf{B}_{\mathrm{S}}^{\top}\mathbf{A}_2^{\top}\hat{\mathbf{C}}_{\mathrm{S}}^{c}\mathbf{A}_1\mathbf{N} + \left(\mathbf{B}_{\mathrm{S}}^{\top}\mathbf{A}_2^{\top}\hat{\mathbf{C}}_{\mathrm{S}}^{c}\mathbf{A}_1\mathbf{N}\right)^{\top} + \mathbf{N}^{\top}\mathbf{A}_1^{\top}\hat{\mathbf{C}}_{\mathrm{S}}^{c}\mathbf{A}_1\mathbf{N} \\ &+ \mathbf{B}_{\mathrm{MB}}^{\top}\left(\mathbf{C}_{\mathrm{MB}}^{\circ} + \mathbf{C}_{\mathrm{MB}}^{\Delta} + \mathbf{C}_{\mathrm{MB}}^{\Delta\top} + \mathbf{A}_3^{\top}\hat{\mathbf{C}}_{\mathrm{M}}^{c}\mathbf{A}_3 + \mathbf{A}_4^{\top}\hat{\mathbf{C}}_{\mathrm{B}}^{c}\mathbf{A}_4\right)\mathbf{B}_{\mathrm{MB}}\Big]\mathbf{a}^{e}\,\mathrm{d}\Omega^{e}\end{aligned} \tag{5.36}$$

In addition to the material properties in $\mathbf{C}_{\square}^{K}$ and $\hat{\mathbf{C}}_{\square}^{K}$ $\forall\square \in \{\mathrm{M}, \mathrm{B}, \mathrm{MB}, \mathrm{S}\} \wedge K \in \{\circ, \Delta, c\}$, the expression now only contains the degrees of freedom $[\mathbf{a}^e]^{\top}$. $\mathbf{A}_i$ $\forall i \in \{2, 3, 4\}$ are auxiliary matrices so that all quantities correspond to the specified vector of degrees of freedom. The detailed structure of all matrices from (5.36) can be found in Appendix C. Equation (4.82) is converted in an analogous way. This results in the following expression.

$$\delta W_{\mathrm{ext}}^{e} = \int\limits_{\partial\Omega_{\mathrm{p}}^{e}} \delta\mathbf{a}^{e\top}\mathbf{N}^{\top}\mathbf{A}_5\mathbf{t}\,\mathrm{d}\,\partial\Omega_{\mathrm{p}}^{e} + \int\limits_{\Omega^e} \delta\mathbf{a}^{e\top}\mathbf{N}^{\top}\mathbf{q}\,\mathrm{d}\Omega^{e} \tag{5.37}$$

The vectors $\mathbf{t}$ and $\mathbf{q}$ contain loads distributed over a curve or surface.

$$\mathbf{t} = \left[\mathbf{n}_\nu^\circ\ \mathbf{n}_\nu^\Delta\ \mathbf{n}_\nu^c\ q_\nu^\circ\ \mathbf{m}_\nu^\circ\ \mathbf{m}_\nu^\Delta\ \mathbf{m}_\nu^c\right]^\top \tag{5.38}$$

$$\mathbf{q} = \left[-s_1\ -s_2\ -s_1\ -s_2\ p\ \tfrac{h^t}{2}s_1\ \tfrac{h^t}{2}s_2\ \tfrac{h^t}{2}s_1\ \tfrac{h^t}{2}s_2\right]^\top \tag{5.39}$$

The vectors $\mathbf{n}_\nu^\circ$, $\mathbf{n}_\nu^\Delta$, $\mathbf{n}_\nu^c$, $\mathbf{m}_\nu^\circ$, $\mathbf{m}_\nu^\Delta$, and $\mathbf{m}_\nu^c$ refer to the Eqs. (4.37)–(4.39), which are given there in tensor notation. Similar to Appendix C, these can be converted into vector-matrix notation and referenced to the element surface Ω^e.

The virtual dynamic work from Eq. (4.83) is represented in vector-matrix notation as follows.

$$\delta W_{\mathrm{dyn}}^e = \int_{\Omega^e} \left[\delta\mathbf{a}^e\right]^\top \mathbf{N}^\top \mathbf{H}\, \mathbf{N}\, \ddot{\mathbf{a}}^e \,\mathrm{d}\Omega^e \tag{5.40}$$

Here $\mathbf{H}$ is an auxiliary matrix and $\ddot{\mathbf{a}}$ are the accelerations. The auxiliary matrix contains global quantities for mass densities and geometries of the layers. Details can be found in Appendix C.

Substituting the Eqs. (5.36), (5.37) and (5.40) into (5.3) results in the equation of motion.

$$\mathbf{M}^e\, \ddot{\mathbf{a}}^e + \mathbf{K}^e\, \mathbf{a}^e = \mathbf{r}^e \tag{5.41}$$

When structuring the term $[\mathbf{a}^e]^\top\, \mathbf{K}^e\, \mathbf{a}^e$ with regard to $\delta\mathbf{a}^{e\top}$ by comparing coefficients, the stiffness matrices for the superposed membrane and the transverse shear state can be determined separately.

$$\mathbf{K}_{\mathrm{MB}}^e = \int_{\Omega^e} \mathbf{B}_{\mathrm{MB}}^\top \left(\mathbf{C}_{\mathrm{MB}}^\circ + \mathbf{C}_{\mathrm{MB}}^\Delta + \mathbf{C}_{\mathrm{MB}}^{\Delta\,\top} + \mathbf{A}_3^\top \hat{\mathbf{C}}_{\mathrm{M}}^c \mathbf{A}_3 + \mathbf{A}_4^\top \hat{\mathbf{C}}_{\mathrm{B}}^c \mathbf{A}_4\right) \mathbf{B}_{\mathrm{MB}} \,\mathrm{d}\Omega^e \tag{5.42}$$

$$\begin{aligned}\mathbf{K}_{\mathrm{S}}^e = \int_{\Omega^e} \Big[& \mathbf{B}_{\mathrm{S}}^\top \left(\mathbf{C}_{\mathrm{S}}^\circ + \mathbf{C}_{\mathrm{S}}^\Delta + \mathbf{C}_{\mathrm{S}}^{\Delta\,\top} + \mathbf{A}_2^\top \hat{\mathbf{C}}_{\mathrm{S}}^c \mathbf{A}_2\right) \mathbf{B}_{\mathrm{S}} \\ & + \mathbf{B}_{\mathrm{S}}^\top \mathbf{A}_2^\top \hat{\mathbf{C}}_{\mathrm{S}}^c \mathbf{A}_1 \mathbf{N} + \left(\mathbf{B}_{\mathrm{S}}^\top \mathbf{A}_2^\top \hat{\mathbf{C}}_{\mathrm{S}}^c \mathbf{A}_1 \mathbf{N}\right)^\top + \mathbf{N}^\top \mathbf{A}_1^\top \hat{\mathbf{C}}_{\mathrm{S}}^c \mathbf{A}_1 \mathbf{N} \Big] \mathrm{d}\Omega^e\end{aligned} \tag{5.43}$$

As already mentioned, the differentiated treatment of the transverse shear stiffness matrix serves the analogous, element-wise, numerical integration. Details can be found in Appendix D. As a result, the merged element stiffness matrix can be determined as follows.

$$\mathbf{K}^e = \mathbf{K}_{\mathrm{MB}}^e + \mathbf{K}_{\mathrm{S}}^e \tag{5.44}$$

The load vector $\mathbf{r}^e$ can be determined from the right-hand side of Eq. (5.41). It includes line loads $\mathbf{r}_1^e$ and area loads $\mathbf{r}_2^e$ on the element.

$$\mathbf{r}_1^e = \int\limits_{\partial\Omega_\mathrm{p}^e} \mathbf{N}^\top \mathbf{A}_5 \mathbf{t} \, \mathrm{d}\,\partial\Omega_\mathrm{p}^e \qquad \mathbf{r}_2^e = \int\limits_{\Omega_\mathrm{p}^e} \mathbf{N}^\top \mathbf{q} \, \mathrm{d}\Omega_\mathrm{p}^e \tag{5.45}$$

The total load vector is composed additively.

$$\mathbf{r}^e = \mathbf{r}_1^e + \mathbf{r}_2^e \tag{5.46}$$

Since in this work primarily areal loads are of interest, a focus on the numerical implementation of this kind of loading should take place here.

For the numerical consideration, the load vector $\mathbf{q}$ is approximated by means of the shape functions.

$$\mathbf{q}(\boldsymbol{\xi}) \approx \mathbf{N}(\boldsymbol{\xi}) \, \mathbf{q}^e \tag{5.47}$$

The element load vector $\mathbf{q}^e$ comprises all nodal load vectors $\mathbf{q}^i$

$$\mathbf{q}^e = \left[\mathbf{q}^1 \; \mathbf{q}^2 \ldots \mathbf{q}^{NN}\right]^\top \tag{5.48}$$

$$\mathbf{q}^i = \left[-s_1^i \; -s_2^i \; -s_1^i \; -s_2^i \; p^i \; \tfrac{h^t}{2} s_1^i \; \tfrac{h^t}{2} s_2^i \; \tfrac{h^t}{2} s_1^i \; \tfrac{h^t}{2} s_2^i\right]^\top \tag{5.49}$$

The variables s_1^i, s_2^i, and p^i are determined by evaluating the load distribution function at the corresponding nodal coordinates $\left(X_1^i, X_2^i\right)$.

$$s_1^i = s_1\left(X_1^i, X_2^i\right) \qquad s_2^i = s_2\left(X_1^i, X_2^i\right) \qquad p^i = p\left(X_1^i, X_2^i\right) \tag{5.50}$$

If Eq. (5.47) is used in Eq. (5.45), the area loads can be approximated as follows.

$$\mathbf{r}_2^e \approx \int\limits_{\Omega_\mathrm{p}^e} \mathbf{N}(\boldsymbol{\xi})^\top \mathbf{N}(\boldsymbol{\xi}) \, \mathbf{q}^e \, \mathrm{d}\Omega_\mathrm{p}^e \tag{5.51}$$

Analogous procedure as for the determination of the stiffness matrix can be used to determine the mass matrix. The result is the following expression.

$$\mathbf{M}^e = \int\limits_{\Omega^e} \mathbf{N}^\top \mathbf{H} \, \mathbf{N} \, \mathrm{d}\Omega^e \tag{5.52}$$

Appendix D is devoted to the numerical integration needed to determine the quantities shown here.

5.8 Assembly

To obtain quantities of the entire structure, all elements $e \in [1, NE]$ in Ω are summed. The direct addition of corresponding element quantities to structure quantities is based on the virtual work of the system, which, as shown in Eq. (5.3), additively consists of the virtual works of all elements. In consequence, the symbolic formulation of this process should be rendered using the $\bigcup$ operator.

$$\mathbf{K} = \bigcup_{e=1}^{NE} \mathbf{K}^e \qquad \mathbf{M} = \bigcup_{e=1}^{NE} \mathbf{M}^e \tag{5.53}$$

$$\mathbf{a} = \bigcup_{e=1}^{NE} \mathbf{a}^e \qquad \mathbf{r} = \bigcup_{e=1}^{NE} \mathbf{r}^e \tag{5.54}$$

The resulting quantities thus refer to the overall structure, which is why the index e is dropped.

5.9 Equation of Motion

By assembling all quantities of the subdomains to quantities of the overall structure, the spatially approximated weak form can be formulated.

$$\begin{aligned} \delta W_{\text{dyn}} \quad + \delta W_{\text{int}} \quad &= \delta W_{\text{ext}} \\ \delta\mathbf{a} \cdot \mathbf{M}\ddot{\mathbf{a}} + \delta\mathbf{a} \cdot \mathbf{K}\mathbf{a} &= \delta\mathbf{a} \cdot \mathbf{r} \end{aligned} \tag{5.55}$$

By Eq. (5.3) it can be deduced that the sum of the virtual work on structural level must be identical to zero.

$$\delta W \approx \delta\mathbf{a} \cdot \left[\mathbf{M}\ddot{\mathbf{a}} + \mathbf{K}\mathbf{a} - \mathbf{r}\right] = 0 \tag{5.56}$$

For arbitrary virtual degrees of freedom $\delta\mathbf{a}$ one obtains the discrete equation of motion.

$$\mathbf{M}\,\ddot{\mathbf{a}} + \mathbf{K}\,\mathbf{a} = \mathbf{r} \tag{5.57}$$

This is an ordinary differential equation of second order in time. In consequence, this equation should be used to solve time-invariant structural-mechanical problems and eigenvalue problems.

5.9.1 Statics

The static case is indicated by $\ddot{\mathbf{a}} = \mathbf{o}$. In this case, a linear system of equations of the solution vector $\mathbf{a}$ results from Eq. (5.57). This system of equations can be stated as follows.

$$\mathbf{K}\,\mathbf{a} = \mathbf{r} \tag{5.58}$$

The solution of this system of equations is realized by the left-hand multiplication with the inverse of the stiffness matrix $\mathbf{K}^{-1}$.

$$\mathbf{K}^{-1}\mathbf{K}\,\mathbf{a} = \mathbf{I}\,\mathbf{a} = \mathbf{a} = \mathbf{K}^{-1}\,\mathbf{r} \tag{5.59}$$

The stiffness matrix must not be singular, since the invertibility is then no longer guaranteed. In order to prevent this, so many Dirichlet boundary conditions must be introduced into the structural equation that no rigid body motions are possible.

5.9.2 Eigenvalue Problem

For eigenvalue problems, only the homogeneous part of the equation of motion is considered. Equation (5.57) is simplified as follows.

$$\mathbf{M}\,\ddot{\mathbf{a}} + \mathbf{K}\,\mathbf{a} = \mathbf{o} \tag{5.60}$$

The solution of the eigenvalue problem is then based on a ansatz for $\mathbf{a}$ and its time derivatives.

$$\begin{aligned} \mathbf{a} &= \boldsymbol{\phi}\,\mathrm{e}^{\imath\omega t} \\ \dot{\mathbf{a}} &= \imath\,\omega\,\boldsymbol{\phi}\,\mathrm{e}^{\imath\omega t} \qquad\qquad \imath^2 = -1 \\ \ddot{\mathbf{a}} &= -\omega^2\boldsymbol{\phi}\,\mathrm{e}^{\imath\omega t} \end{aligned}$$

Here $\boldsymbol{\phi}$ is the eigenvector, ω is the eigenvalue, $\imath$ is the imaginary part of the complex number and t is the time. By substitution of the vector of the degrees of freedom and its time derivative with the above ansatz, the equation of motion for (5.60) results as follows.

$$-\omega^2\mathbf{M}\,\boldsymbol{\phi}\,\mathrm{e}^{\imath\omega t} + \mathbf{K}\,\boldsymbol{\phi}\,\mathrm{e}^{\imath\omega t} = \mathbf{o} \tag{5.61}$$

The reformulation yields the generalized eigenvalue problem.

$$\left[\mathbf{K} - \omega^2\mathbf{M}\right]\boldsymbol{\phi} = \mathbf{o} \tag{5.62}$$

Since the generalized eigenvalue problem described above works efficiently only for small mechanical systems, the special eigenvalue problem is used here. The transfer into the special eigenvalue problem occurs through the factorization of the mass matrix in accordance to Cholesky [4].

$$\mathbf{M} = \mathbf{L}\,\mathbf{L}^{\top} \tag{5.63}$$

Here $\mathbf{L}$ is a lower triangular matrix. By substituting this decomposition into the generalized eigenvalue problem, the following equation results.

$$\mathbf{K}\boldsymbol{\phi} - \omega^2\mathbf{L}\mathbf{L}^{\top}\boldsymbol{\phi} = \mathbf{K}\mathbf{L}^{-1}\boldsymbol{\Phi} - \omega^2\mathbf{L}\boldsymbol{\Phi} = \mathbf{L}^{-1}\mathbf{K}\mathbf{L}^{-\top}\boldsymbol{\Phi} - \omega^2\mathbf{I}\boldsymbol{\Phi} = \mathbf{o} \tag{5.64}$$

The special eigenvalue problem can thus be formulated as follows.

$$\left[\mathbf{L}^{-1}\mathbf{K}\,\mathbf{L}^{-\top} - \omega^2\,\mathbf{I}\right]\boldsymbol{\Phi} = \mathbf{o} \tag{5.65}$$

The solution is done by determining the eigenvalues ω_i using the non-trivial solution.

$$\left[\mathbf{L}^{-1}\mathbf{K}\,\mathbf{L}^{-\top} - \omega^2\,\mathbf{I}\right] = \mathbf{o} \tag{5.66}$$

The eigenvector of the special eigenvalue problem can be determined by substituting the eigenvalues in Eq. (5.65). The back transformation of the eigenvector is done as follows.

$$\boldsymbol{\phi} = \mathbf{L}^{-\top}\,\boldsymbol{\Phi} \tag{5.67}$$

For specific solvers we refer to the technical literature, e.g. [1]. For the subsequent computations, the Lanczos iteration method is applied [5].

5.10 Implementation with ABAQUS

For the numerical solution of the different problems the commercial program system ABAQUS (version 6.14) is used. To solve the equations described in this chapter, an user-defined element subroutine (UEL) based on the work presented in [3] was created in FORTRAN (`.for`). The extension concerns the definition of all stress resultants as well as the global degrees of freedom as solution-dependent variables (SDV). However, the program system no longer supports graphical input when using user-defined elements. Modeling via ABAQUS/CAE is therefore not possible. For efficient generation of input files (`.inp`), which can be read by the program system via a command line interface, a MATLAB script was generated. In addition to the variation of the input parameters, it is just as easy to perform discretization variations. In this input file one can decide between static calculations and eigenvalue analyzes. The basic process of a calculation is shown schematically in Fig. 5.3. The processes

input file	.inp
geometry	$L_\alpha^K, h^K \ \forall K = \{t,c,b\}$
material	$Y^K, \nu^K, \rho^K \ \forall K = \{t,c,b\}$
loading	$\{s_1(\boldsymbol{r}), s_2(\boldsymbol{r}), p(\boldsymbol{r})\}|_{\Gamma_{\mathrm N}}$
support	$\{v_\alpha^K, \varphi_\alpha^K, w\}|_{\Gamma_{\mathrm D}} \ \forall K = \{t,c,b\}$
FE-topology for UEL and visualizing elements	
integration scheme	full, reduced, selective

FORTRAN file	.for
UEL	
definition finites element	$\mathit{PG}, \mathbf{N}, \mathit{GP}$
definition solution variables	$v_\alpha^K, \varphi_\alpha^K, w, N_{\alpha\beta}^K, Q_\alpha^K, M_{\alpha\beta}^K, G_{\alpha\beta}^K, g_\alpha^K, K_{\alpha\beta}^K \ \forall K = \{\circ, \Delta, c\}$
element stiffness matrix membrane-bending portion	$\mathbf{K}_{\mathrm{MB}}^e$
kinemtic matrix membrane-bending portion	$\mathbf{B}_{\mathrm{MB}}^e$
element stiffness matrix transverse shear portion	$\mathbf{K}_{\mathrm S}^e$
kinemtic matrix transverse shear portion	$\mathbf{B}_{\mathrm S}^e$
element mass matrix	$\mathbf{M}^e$
right-hand side vector	$\mathbf{r}^e$
UMAT (visualization purposes)	
transmission of solution variables to visualization element	

ABAQUS solver	
assembly and computation	
static loading	$\mathbf{K}\,\mathbf{a} = \mathbf{r}$
eigenvalue problem	$\left[\mathbf{K} - \omega^2 \mathbf{M}\right] \boldsymbol{\phi} = \mathbf{o}$

output	
results database	.odb
further files	.res, .msg, etc.

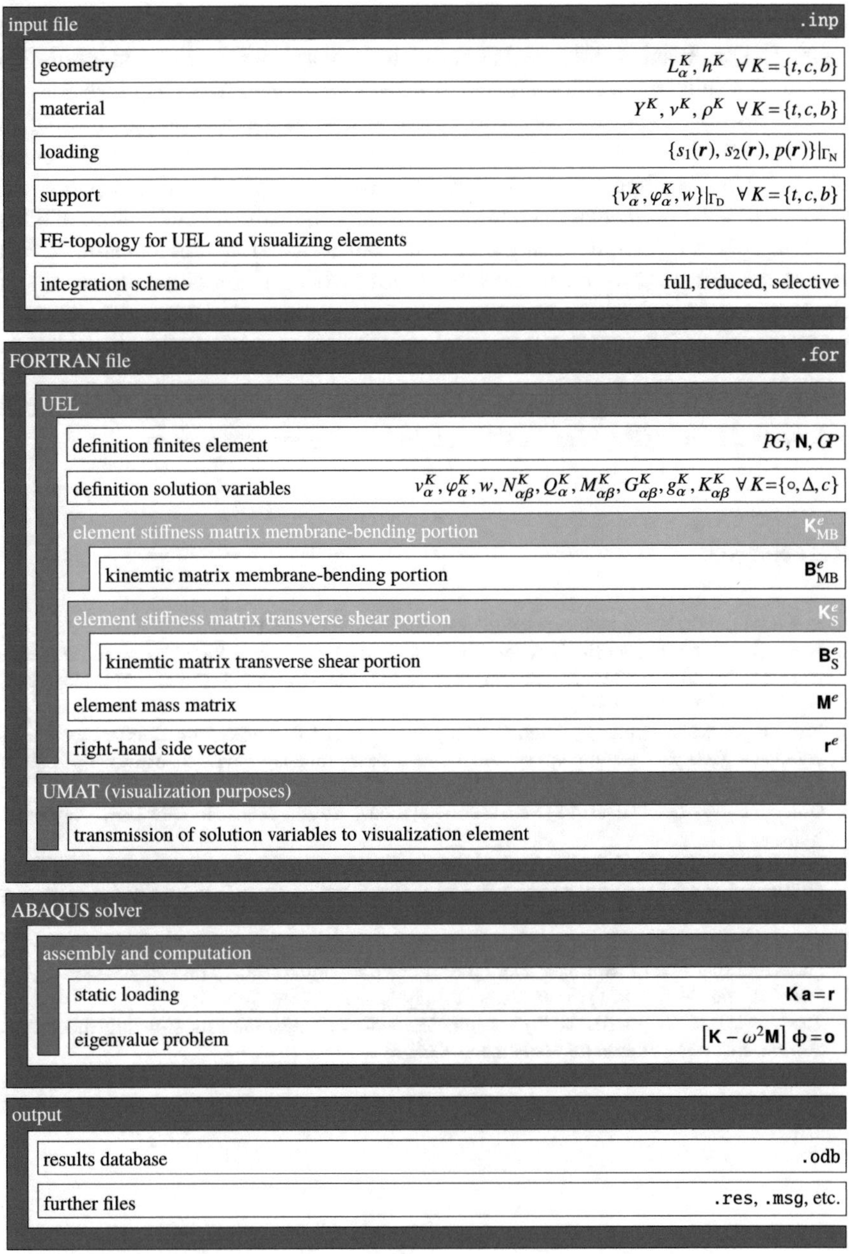

Fig. 5.3 Structure for integrating the various routines in the FE program sequence

running in the ABAQUS solver are handled as 'black box' in the numerical treatment. The preceding section deals only with basic approaches to the solution. The generated results can not be evaluated in ABAQUS/Viewer. ABAQUS also does not provide result visualization for custom elements, as the number and location of the Gauss points of the custom element are unknown. Therefore, a flat default element of the ABAQUS library with identical location and number of Gauss points was used, positioned on matching coordinates of the original structure's computational elements, and the post-processing results are transferred to the default element. This was implemented in a user material subroutine (UMAT). The additionally introduced elements possess no rigidity. They thus do not affect the result. Only result quantities are transmitted. They are therefore used only for color-coded representation of the results.

All calculations are performed on a computer with a quad core Intel i7-3820 processor, 32 GB of RAM and a 64-bit Windows 7 operating system.

References

1. Bathe KJ (2002) Finite-elemente-methoden, 2nd edn. Springer, Berlin
2. Bathe KJ, Dvorkin EN (1985) A four-node plate bending element based on mindlin/reissner plate theory and a mixed interpolation. Int J Numer Methods Eng 21(2):367–383. https://doi.org/10.1002/nme.1620210213
3. Eisenträger J, Naumenko K, Altenbach H, Meenen J (2015) A user-defined finite element for laminated glass panels and photovoltaic modules based on a layer-wise theory. Compos Struct 133:265–277. https://doi.org/10.1016/j.compstruct.2015.07.049
4. Golub GH, Van Loan CF (1996) Matrix computations, 3rd edn. Johns Hopkins University Press, Baltimore
5. Lanczos C (1950) An iteration method for the solutions of the eigenvalue problem of linear differential and integral operators. J Res Natl Bur Stand 45(4):255–282. https://nvlpubs.nist.gov/nistpubs/jres/045/4/V45.N04.A01.pdf
6. Oñate E (2013) Structural analysis with the finite element method linear statics: volume 2. Beams, plates and shells, vol 2. Springer, Dordrecht. https://doi.org/10.1007/978-1-4020-8743-1_6
7. Szabó B, Babuška I (1991) Finite element analysis. Wiley, New York
8. Zienkiewicz OC, Taylor RL (2005) The finite element method for solid and structural mechanics, 6th edn. Elsevier Butterworth-Heinemann, Oxford

Chapter 6
Convergence and Verification

6.1 Model Problem

At this point, the convergence of the numerical solution is to be checked and the numerical solution to be verified with a closed-form solution based on two limiting cases. Ultimately, the suitability of the finite element can be proven. It is based on a simple load and support situation. As illustrated in Fig. 6.1 on the left-hand side, a homogeneous and orthogonal load is used as model problem for all considerations in this chapter. For the sake of simplicity, the structure to be examined is symmetric in the transverse direction. The cover layers thus have identical geometric dimensions and material properties. Regarding supports, only the deflections at the edges with normal vector $\boldsymbol{\nu}$ are prohibited. These edges are thus supported torque-free. The geometric dimensions and material characteristics are exemplary.

Next, evaluations of convergence for the static loading and the eigenvalue problem are made. In this context, the computation times are also discussed. As a result, the classical structural theories for thin-walled members for the shear-rigid and the shear soft case are used to generate closed-form solutions while the geometrical and material parameters structure of the numerical procedure is adapted according to the assumptions of these theories. Afterwards, the computational results are compared with the ones of the closed-form solutions.

6.2 Convergence Analysis

Within the framework of the convergence analysis, the required number of finite elements of the spatial discretization is to be determined by means of h^e-adaption. As introduced in Chap. 5, it is restricted to the polynomial degree $PG = 2$. In addition, the convergence behavior of three types of integration is compared (full, reduced, selective). As a restriction, $h^e_\alpha = h^e_\beta$ is introduced for the elements edge lengths. Thus, the aspect ratio in the entire domain is $AR = {}^{h^e_{\max}}/_{h^e_{\min}} = 1$ for all discretization

M. Aßmus, *Structural Mechanics of Anti-Sandwiches*,
SpringerBriefs in Continuum Mechanics,
https://doi.org/10.1007/978-3-030-04354-4_6

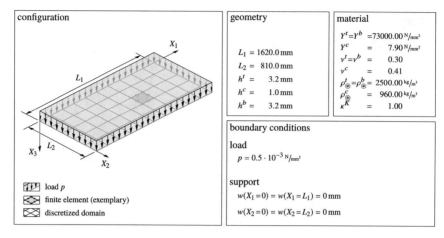

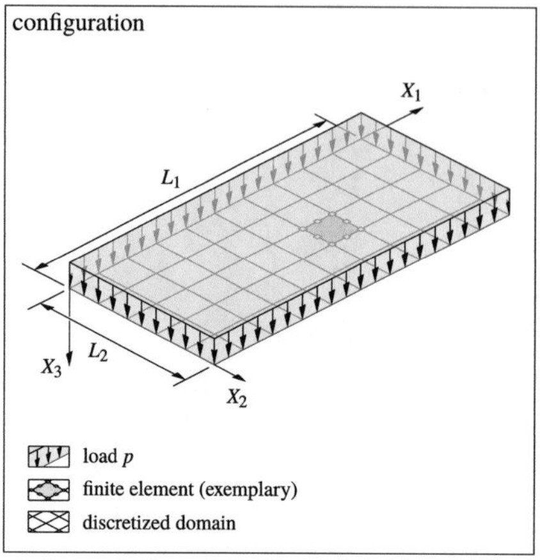

geometry

$L_1 = 1620.0\,\mathrm{mm}$
$L_2 = 810.0\,\mathrm{mm}$
$h^t = 3.2\,\mathrm{mm}$
$h^c = 1.0\,\mathrm{mm}$
$h^b = 3.2\,\mathrm{mm}$

material

$Y^t = Y^b = 73000.00\,\mathrm{N/mm^2}$
$Y^c = 7.90\,\mathrm{N/mm^2}$
$\nu^t = \nu^b = 0.30$
$\nu^c = 0.41$
$\rho^t_{\circledast} = \rho^b_{\circledast} = 2500.00\,\mathrm{kg/m^3}$
$\rho^c_{\circledast} = 960.00\,\mathrm{kg/m^3}$
$\kappa^K = 1.00$

boundary conditions

load
$p = 0.5 \cdot 10^{-3}\,\mathrm{N/mm^2}$

support
$w(X_1 = 0) = w(X_1 = L_1) = 0\,\mathrm{mm}$
$w(X_2 = 0) = w(X_2 = L_2) = 0\,\mathrm{mm}$

Fig. 6.1 Configuration, geometry, material and boundary conditions for convergence analysis

variants. A structured, undistorted mesh it used where all element inner angles are 90°.

The discretization variants introduced are documented in Table 6.1. The respective number of elements (NE), nodes (NN), Gauss points (NG) and system degrees of freedom (NEQ) are specified there. The mesh density increases with the variant number. The introduced elements with quadratic shape functions (8 nodes per element) are used. The specified number of Gauss points refers to full integration with 9 integration points per element. The number of degrees of freedom NEQ refers to the nodal degrees of freedom introduced for numerics from Eq. (5.6) ($NDOF = 9$). The following relationships apply to the two dimensional discretization if the element introduced in Chap. 5 is used.

$$NE(\Omega) = NE(\boldsymbol{e}_1) \cdot NE(\boldsymbol{e}_2) \tag{6.1}$$

$$NN(\Omega) = [2NE(\boldsymbol{e}_1) + 1]\,[NE(\boldsymbol{e}_2) + 1] + [NE(\boldsymbol{e}_1) + 1]\,NE(\boldsymbol{e}_2) \tag{6.2}$$

$$NG(\Omega) = NE(\Omega) \cdot NG(\Omega^e) \tag{6.3}$$

$$NEQ = NN(\Omega) \cdot NDOF \tag{6.4}$$

Table 6.1 Characteristics of the discretization variation for the model problem

variant	$NE(\boldsymbol{e}_1)$	$NE(\boldsymbol{e}_2)$	$NE(\Omega)$	$NN(\Omega)$	$NG(\Omega)$	NEQ
1	8	4	32	121	288	1089
2	16	8	128	433	1152	3897
3	32	16	512	1633	4608	14697
4	64	32	2048	6337	18432	57033
5	128	64	8192	24961	73728	224649

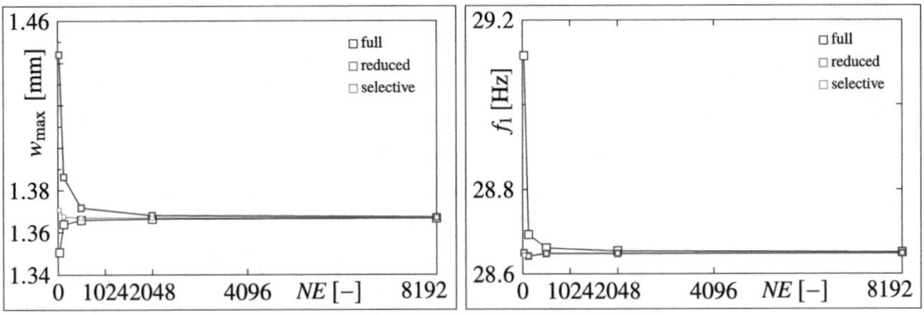

Fig. 6.2 Convergence of maximum deflection w_{max} and first eigenfrequency f_1

As a simple criterion for verifying convergence, the maximum deflection is used in the static load investigations. The convergence criterion is the achievement of a steady state value. Since the load in the present case is applied orthogonally and uniformly, the maximum deflection is $w_{\mathrm{max}} = w(L_1/2, L_2/2)$. In order to remain within the scope of geometric linear theory ($w_{\mathrm{max}} \leq 0,5H$), a relatively small load was chosen. As shown in the results in Fig. 6.2 on the left-hand side, this limit is undercut ($H = 7,4\,\mathrm{mm}$).

When solving the eigenvalue problem, the smallest eigenfrequency is used as the result variable for the convergence evaluation. The problem is considered convergent when this frequency approaches a limit. The element edge length of a finite element should be significantly smaller than the minimum wavelength to be examined in dealing with eigenvalue problems so that the element shape functions can adequately resolve the waves. The following relationship applies between element edge length and minimum wavelength λ_α.

$$\frac{\lambda_\alpha}{h^e_\alpha} \geq 4 \tag{6.5}$$

If, in the given structure it is assumed that a maximum of three half-waves of each structure edge length for the highest natural frequency is of interest, and considering the factor 5 in Eq. (6.5) for the sake of security, a minimum required element edge length of $h^e_\alpha = 54\,\mathrm{mm}$ for the short edge $L_2 = 810\,\mathrm{mm}$ results.

6.2.1 Discussion

In the evaluations, we only refer to the maximum deflections and the first eigenfrequency. For both investigations it becomes obvious that the problem is already solved with sufficient accuracy when utilizing 2048 elements. The quadrupling of the number of elements improves the result only slightly. For complete integration these are $\Delta w_{\mathrm{max}} = 2.3 \cdot 10^{-4}\,\mathrm{mm}$ or $\Delta f = 3 \cdot 10^{-3}\,\mathrm{Hz}$. It turns out that selective integration

Table 6.2 Computation times $t_\square$ $\forall \square \in \{\mathrm{f}, \mathrm{r}, \mathrm{s}\}$ at discretization- and integration variation

	Static loading			Eigenfrequency analysis		
NE [−]	t_f (s)	t_r (s)	t_s (s)	t_f (s)	t_r (s)	t_s (s)
32	0.4	0.2	0.3	0.5	0.3	0.5
128	1.3	0.7	1.0	1.5	1.0	1.4
512	4.6	2.4	3.5	6.0	3.4	5.1
2048	22.7	9.6	14.7	25.6	13.5	19.1
8192	75.5	48.6	61.0	115.5	58.9	84.4

the best convergence behavior. However, reduced and selective integration converge unphysical due to the non-exactness of the reduced-order integration.

In the analyzes of the eigenvalue problem, the computation times were determined in quest of the first twelve eigenfrequencies. The results are detailed in Table 6.2 for both problems. It turns out that the reduced integration provides the shortest computation times. By contrast, the use of full integration results in the longest computation periods. The results are also visualized in Fig. 6.3. The computation periods were normalized with the respectively used number of elements. Result is the average computation time per element. It turns out that for values $NE = 512 \ldots 2048$ a minimum can be found for all integration types.

6.3 Verification with Monolithic Solutions

After checking the convergence of the finite element for typical material and geometry data, the question remains whether the developed equations are solved correctly. The verification of the suitability of the finite element is to be done on the basis of the thin-walled structural theories of Kirchhoff and Mindlin. While the former theory considers shear-rigid structures, the latter is related to shear soft structures. Both theories follow the bending problem. The membrane problem is completely

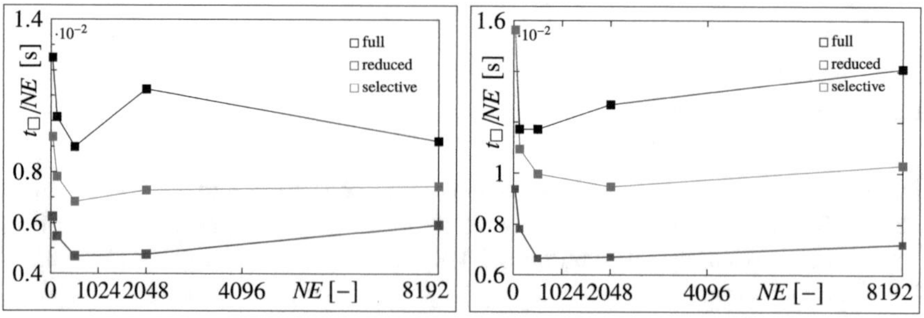

Fig. 6.3 Normalized computation times at static loading (left) and eigenfrequency analysis (right)

neglected. It becomes obvious that both theories mentioned above are special cases of the theory presented in Chaps. 2 or 3, respectively. Furthermore, the shear-rigid theory is a special case of the shear soft theory. If the shear stiffness of a thin-walled structural element increases towards infinity, the shear-rigid problem results.

The numerical solutions are now to be conducted on the basis of the equalization of the material properties of all layers ($Y = Y^t = Y^c = Y^b$ and $\nu = \nu^t = \nu^c = \nu^b$). It now becomes apparent that two limiting cases arise. If Young's modulus and Poisson's ratio of all layers are equated with those of the skin layers (for values, see Fig. 6.1), the result is the shear-rigid case. However, if the considerations are based on the material properties of the core layer, the shear soft case arises.

The approach presented here is legitimate and sufficient to verify the functionality. In contrast, there are solutions in closed form, as e.g. reproduced in [2]. Nevertheless, at this point, such a strategy is dispensed with, since results obtained with it are also based on a special case only while the boundary conditions introduced there cause problems lacking physical interpretation.

The deviation of the computational (comp) to the closed-form solution (cf) is determined according to the case under investigation as follows by means of absolute (Δ) and relative (δ) error measures. In the sequel we reduce the verification to results of the maximum deflection and the first eigenfrequency.

$$\Delta_{w_{\max}} = w_{\max}^{\text{comp}} - w_{\max}^{\text{cf}} \qquad \delta_{w_{\max}} = \frac{\Delta_{w_{\max}}}{w_{\max}^{\text{cf}}} \qquad \forall\, w_{\max}^{\text{cf}} \neq 0 \tag{6.6}$$

$$\Delta_f = f^{\text{comp}} - f^{\text{cf}} \qquad \delta_f = \frac{\Delta_f}{f^{\text{cf}}} \qquad \forall\, f^{\text{cf}} \neq 0 \tag{6.7}$$

6.3.1 Shear-Rigid Plate

The shear rigid problem for planar surfaces can be solved using the Kirchhoff theory [4]. Basically, the Kirchhoff theory can be derived from the theory of shear soft plates emanate from Mindlin or Reissner, whose basic features have been used so far. It is assumed that the shear stiffness tends towards infinity. An orthogonal straight line to the mid surface before the deformation remains orthogonal and straight during the deformation process. Only orthogonal loads are considered. Without further derivation, the basic equations of the Kirchhoff theory are reproduced below. The following differential equation serves as a basis.

$$D_{\text{B}} \nabla^2 \nabla^2 w + \rho_{\circledast} H \ddot{w} = p \tag{6.8}$$

Here D_{B} is the plate stiffness from Eq. (B.15). The change of sign of load and inertia terms must be observed. This inhomogeneous partial differential equation of the 4th order for the homogeneous, shear-rigid plate will serve as the basis for the verification of the computational solution. For this purpose, two cases are examined, both of which are based on an all-round hinged plate.

Neglecting the term of inertia ($\rho_{\circledast} H \ddot{w} = 0$) in Eq. (6.8), the following equation is used to analyze the plate deflection.

$$D_{\mathrm{B}} \nabla^2 \nabla^2 w = p \tag{6.9}$$

To solve such equations one uses trigonometric series. The solution method chosen here is the Navier double-series solution approach [7]. For this, the deflections are developed in a double sine series.

$$w(\boldsymbol{r}) = \sum_{m=1}^{\infty} \sum_{n=1}^{\infty} w_{mn} \sin\left(\frac{m\pi}{L_1} X_1\right) \sin\left(\frac{n\pi}{L_2} X_2\right) \tag{6.10}$$

Here w_{mn} is the maximum deflection and m, n are control variables. The deflections should be hindered at all edges. Also, all edges should be free of moments.

$$w \quad (0, X_2) = w \quad (L_1, X_2) = 0 \qquad w \quad (X_1, 0) = w \quad (X_1, L_2) = 0 \tag{6.11}$$

$$M_{11}(0, X_2) = M_{11}(L_1, X_2) = 0 \qquad M_{22}(X_1, 0) = M_{22}(X_1, L_2) = 0 \tag{6.12}$$

The solution (6.10) fulfills the required boundary conditions. The derivation of this approach leads to the following expression.

$$\nabla^2 \nabla^2 w = \sum_{m=1}^{\infty} \sum_{n=1}^{\infty} \left[\left(\frac{m\pi}{L_1}\right)^4 + 2\left(\frac{m\pi}{L_1}\right)^2 \left(\frac{n\pi}{L_2}\right)^2 + \left(\frac{n\pi}{L_2}\right)^4\right] w_{mn} \sin\left(\frac{m\pi}{L_1} X_1\right) \sin\left(\frac{n\pi}{L_2} X_2\right) \tag{6.13}$$

The load should act orthogonally and homogeneously on the plate surface. This, too, can be developed in a double sine series.

$$p(\boldsymbol{r}) = \sum_{m=1}^{\infty} \sum_{n=1}^{\infty} p_{mn} \sin\left(\frac{m\pi}{L_1} X_1\right) \sin\left(\frac{n\pi}{L_2} X_2\right) \tag{6.14}$$

Herein p_{mn} is the load amplitude, which can be stated in general form as follows [8].

$$p_{mn} = \frac{4}{L_1 L_2} \int_0^{L_1} \int_0^{L_2} p(X_1, X_2) \sin\left(\frac{m\pi}{L_1} X_1\right) \sin\left(\frac{n\pi}{L_2} X_2\right) \, \mathrm{d}X_1 \, \mathrm{d}X_2 \tag{6.15}$$

For a homogeneous load with $p(X_1, X_2) = p_0$ we obtain the following Fourier coefficients [1].

$$p_{mn} = \frac{16 q_0}{\pi^2 mn} \qquad \forall \, m, n \in \mathbb{N}_{\star}^{+} \tag{6.16}$$

Substituting into Eq. (6.9) results in the following relation.

$$\sum_{m=1}^{\infty}\sum_{n=1}^{\infty}\left[\left(\frac{m\pi}{L_1}\right)^4+2\left(\frac{m\pi}{L_1}\right)^2\left(\frac{n\pi}{L_2}\right)^2+\left(\frac{n\pi}{L_2}\right)^4\right]w_{mn}\sin\left(\frac{m\pi}{L_1}X_1\right)\sin\left(\frac{n\pi}{L_2}X_2\right)$$
$$=\frac{1}{D_\mathrm{B}}\sum_{m=1}^{\infty}\sum_{n=1}^{\infty}p_{mn}\sin\left(\frac{m\pi}{L_1}X_1\right)\sin\left(\frac{n\pi}{L_2}X_2\right) \tag{6.17}$$

By comparing coefficients and attending the binomial, the maximum deflection of the plate can be expressed as compact as follows.

$$w_{mn}=\frac{p_{mn}}{D_\mathrm{B}}\left[\left(\frac{m\pi}{L_1}\right)^2+\left(\frac{n\pi}{L_2}\right)^2\right]^{-2} \tag{6.18}$$

The insertion in the Navier approach provides the location-dependent deflection.

$$w(\boldsymbol{r})=\sum_{m=1}^{\infty}\sum_{n=1}^{\infty}\frac{p_{mn}}{D_\mathrm{B}}\left[\left(\frac{m\pi}{L_1}\right)^2+\left(\frac{n\pi}{L_2}\right)^2\right]^{-2}\sin\left(\frac{m\pi}{L_1}X_1\right)\sin\left(\frac{n\pi}{L_2}X_2\right)\quad \forall m,n\in\mathbb{N}_\star^+ \tag{6.19}$$

In Table 6.3 the closed-form solution of the plate deflection is compared with the computational solution obtained with XLWT. The evaluation is limited to the maximum deflection. Again, $w_{\max}=w(^{L_1}/_2, {}^{L_2}/_2)$ holds true. The deviation is vanishingly small.

The starting point for the eigenvalue analysis is the undamped shear-rigid elastic plate, whereby in Eq. (6.8) loads are disregarded, but the inertia terms are considered now.

$$D_\mathrm{B}\nabla^2\nabla^2 w+\rho_{\circledast}H\ddot{w}=0 \tag{6.20}$$

First, the Bernoulli ansatz [3] is introduced to separate variables of space and time.

$$w(\boldsymbol{r},t)=W(\boldsymbol{r})\,T(t) \tag{6.21}$$

The application of this approach leads to the following equation.

Table 6.3 Resulting maximum deflection of the shear-rigid solutions in comparison

Closed-form (Kirchhoff)	Computational (XLWT)	Deviation	
$w_{\max}$ (mm)	$w_{\max}$ (mm)	$\Delta_{w_{\max}}$ (mm)	$\delta_{w_{\max}}$ (%)
$16.111\cdot10^{-3}$	$16.128\cdot10^{-3}$	$1.7\cdot10^{-5}$	0, 11

$$\frac{D_B \nabla^2 \nabla^2 W(\boldsymbol{r})}{\rho_{\circledast} H W(\boldsymbol{r})} = -\frac{\ddot{T}(t)}{T(t)} = \omega^2 \tag{6.22}$$

This gives the following two equations.

$$\nabla^2 \nabla^2 W(\boldsymbol{r}) - \lambda^4 W(\boldsymbol{r}) = 0 \qquad \text{with } \lambda^4 = \frac{\rho_{\circledast} H \omega^2}{D_B} \tag{6.23}$$

$$\ddot{T}(t) + \omega^2 T(t) = 0 \tag{6.24}$$

The time dependence is represented by a harmonic function $T = A \sin(\omega t + \alpha)$. Herein, α is a constant phase shift. The following approach captures the time dependence and gives the eigenvalue problem for the plane consideration.

$$w(\boldsymbol{r}, t) = W(\boldsymbol{r}) \sin(\omega t + \alpha) \tag{6.25}$$

Also at this point, the Navier double-series solution approach for a moment free supported plate is used [1].

$$W(\boldsymbol{r}) = \sum_{m=1}^{M} \sum_{n=1}^{N} W_{mn} \sin\left(\frac{m\pi}{L_1}\right) X_1 \sin\left(\frac{n\pi}{L_2}\right) X_2 \tag{6.26}$$

The resulting eigenmode shapes are shown in Fig. 6.4.

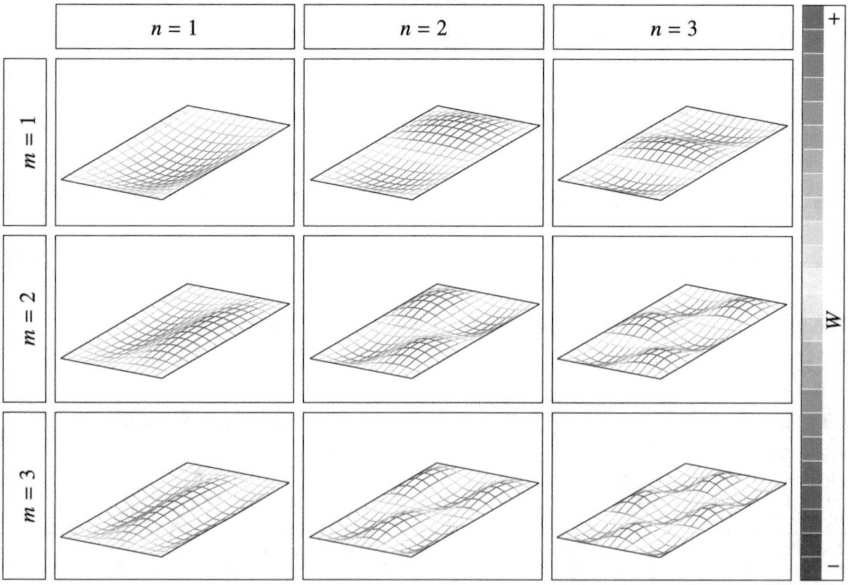

Fig. 6.4 Eigenmode shapes of the shear-rigid plate with free supports for the shear rigid case

After substituting Eq. (6.26) into $\nabla^2\nabla^2 W - \lambda^4 W = 0$ and comparing coefficients, the following equation results.

$$\sum_{m=1}^{M}\sum_{n=1}^{N}\left[W_{mn}\left(\left(\frac{m\pi}{L_1}\right)^2+\left(\frac{n\pi}{L_2}\right)^2\right)^2-\lambda^4\right]\sin\left(\frac{m\pi}{L_1}\right)X_1\sin\left(\frac{n\pi}{L_2}\right)X_2=0 \tag{6.27}$$

This results in the eigenvalue λ as follows.

$$\lambda^4=\pi^4\left(\frac{m^2}{L_1^2}+\frac{n^2}{L_2^2}\right)^2 \qquad \Rightarrow \qquad \frac{\rho_{\circledast} H\omega^2}{D_{\mathrm{B}}}=\pi^4\left(\frac{m^2}{L_1^2}+\frac{n^2}{L_2^2}\right)^2 \tag{6.28}$$

Considering the correlation $\omega = 2\pi f$, the natural frequency f can be determined as follows.

$$f_{mn}=\frac{\pi}{2}\left(\frac{m^2}{L_1^2}+\frac{n^2}{L_2^2}\right)\sqrt{\frac{D_{\mathrm{B}}}{\rho_{\circledast} H}} \qquad \forall\, m,n\in\mathbb{N}^+ \tag{6.29}$$

Resulting analytically determined eigenfrequencies are juxtaposed with the computationally determined in Table 6.4. Absolute values of the deviations of less than 0.5% in the entire investigation are determined.

6.3.2 Shear-Soft Plate

Analogous to the executions in Sect. 6.3.1, the differential equations of the shear soft plate are to be reproduced without derivation.

Table 6.4 Resulting natural frequencies of the shear-rigid solutions in comparison

Closed-form (Kirchhoff)									
mn	11	12	13	21	22	23	31	32	33
f_{mn} (Hz)	36.21	123.12	267.98	57.94	144.85	289.70	94.16	181.07	325.92
Computational (XLWT)									
r	1	2	3	4	5	6	7	8	9
f_r (Hz)	36.14	122.94	267.38	57.77	144.46	288.81	93.88	180.39	324.59
Deviation									
Δ_f (Hz)	−0.07	−0.18	−0.60	−0.17	−0.39	−0.89	−0.28	−0.68	−1.33
δ_f (%)	−0.19	−0.07	−0.22	−0.29	−0.27	−0.31	−0.30	−0.38	−0.41

$$D_S \nabla^2 (\nabla \cdot \boldsymbol{\varphi}) + p = \rho_{\circledast} H \ddot{w} \tag{6.30}$$

$$\nabla^2 w + \nabla \cdot \boldsymbol{\varphi} = \frac{D_B}{D_S} \nabla^2 (\nabla \cdot \boldsymbol{\varphi}) + \frac{\rho_{\circledast} H^3}{12} \ddot{\boldsymbol{\varphi}} \tag{6.31}$$

$$\frac{1-\nu}{2} \frac{D_B}{D_S} \nabla^2 \left[(\nabla \times \boldsymbol{\varphi}) \cdot \boldsymbol{n} \right] = (\nabla \times \boldsymbol{\varphi}) \cdot \boldsymbol{n} + \frac{\rho_{\circledast} H^3}{12} \ddot{\boldsymbol{\varphi}} \tag{6.32}$$

The rotations are treated as independent variables here again. Since the three differential equations are each 2nd order, a 6th order problem results. For simplicity, the following abbreviations can be introduced.

$$\Phi = (\varphi_{1,1} + \varphi_{2,2}) \tag{6.33}$$

$$\Phi_{,1} = (\varphi_{1,11} + \varphi_{2,12}) \tag{6.34}$$

$$\Phi_{,2} = (\varphi_{1,12} + \varphi_{2,22}) \tag{6.35}$$

$$\nabla^2 \varphi_1 = \varphi_{1,11} + \varphi_{1,22} \tag{6.36}$$

$$\nabla^2 \varphi_2 = \varphi_{2,11} + \varphi_{2,22} \tag{6.37}$$

Thus, the Eqs. (6.30), (6.31), and (6.32) can be represented as follows [1].

$$D_S \left(\nabla^2 w + \Phi \right) + p = \rho_{\circledast} H \ddot{w} \tag{6.38}$$

$$\frac{D_B}{2} \left[(1-\nu) \nabla^2 \varphi_1 + (1+\nu) \Phi_{,1} \right] - D_S (\varphi_1 + w_{,1}) = \frac{\rho_{\circledast} H^3}{12} \ddot{\varphi}_1 \tag{6.39}$$

$$\frac{D_B}{2} \left[(1-\nu) \nabla^2 \varphi_2 + (1+\nu) \Phi_{,2} \right] - D_S (\varphi_2 + w_{,2}) = \frac{\rho_{\circledast} H^3}{12} \ddot{\varphi}_2 \tag{6.40}$$

The limit to the shear-rigid plate results from $D_S \to \infty$.

Now the angles φ_α are to be eliminated. Adding Eqs. $(6.39)_{,1}$ and $(6.39)_{,2}$ gives

$$D_B \left(\nabla^2 \Phi \right) - D_S (\Phi + \nabla^2 w) = \frac{\rho_{\circledast} H^3}{12} \ddot{\Phi}\,. \tag{6.41}$$

Considering $\Phi = {(\rho_{\circledast} H \ddot{w} - p)}/{D_S} - \nabla^2 w$ from Eq. (6.38) results in the following expression.

$$\left[\nabla^2 - \frac{\rho_{\circledast} H}{D_S} \frac{\partial^2}{\partial t^2} \right] \left[D_B \nabla^2 - \frac{\rho_{\circledast} H^3}{12} \frac{\partial^2}{\partial t^2} \right] w + \rho_{\circledast} H \frac{\partial^2 w}{\partial t^2}$$
$$= \left[1 - \frac{D_B}{D_S} \nabla^2 + \frac{\rho_{\circledast} H^3}{12 D_S} \frac{\partial^2}{\partial t^2} \right] p \tag{6.42}$$

Thus, a differential equation of 4th order with respect to space and time results.

As with the shear rigid problem in the static considerations, all inertia terms are neglected. The set of equations according to Eqs. (6.38)–(6.40) then results as follows.

$$D_S\left(\nabla^2 w + \Phi\right) + p = 0 \tag{6.43}$$

$$\frac{D_B}{2}\left[(1-\nu)\nabla^2\varphi_1 + (1+\nu)\Phi_{,1}\right] - D_S(\varphi_1 + w_{,1}) = 0 \tag{6.44}$$

$$\frac{D_B}{2}\left[(1-\nu)\nabla^2\varphi_2 + (1+\nu)\Phi_{,2}\right] - D_S(\varphi_2 + w_{,2}) = 0 \tag{6.45}$$

The load can be developed according to the procedure in (6.14)–(6.16). Furthermore, the deflections can be eliminated by introducing the expression $\tilde{w} = w - \frac{D_B}{D_S}\Phi$. This results in the following expression.

$$\nabla^2\Phi = -\frac{p}{D_B} \tag{6.46}$$

Taking into account the series expansion, the solution is as follows.

$$\Phi = \frac{1}{D_B}\sum_{m=1}^{\infty}\sum_{n=1}^{\infty} p_{mn}\left[\left(\frac{m\pi}{L_1}\right)^2 + \left(\frac{n\pi}{L_2}\right)^2\right]\sin\left(\frac{m\pi}{L_1}X_1\right)\sin\left(\frac{n\pi}{L_2}X_2\right) \tag{6.47}$$

This solution is then inserted into Eq. (6.43) and rearranging to $\nabla^2 w$.

$$\nabla^2 w = -\sum_{m=1}^{\infty}\sum_{n=1}^{\infty}\left\{\frac{p_{mn}}{D_B}\left[\left(\frac{m\pi}{L_1}\right)^2 + \left(\frac{n\pi}{L_2}\right)^2\right]^{-1} + \frac{p_{mn}}{D_S}\right\}\sin\left(\frac{m\pi}{L_1}X_1\right)\sin\left(\frac{n\pi}{L_2}X_2\right) \tag{6.48}$$

Integration can now be used to determine the location-dependent deflections.

$$w(\boldsymbol{r}) = \frac{1}{D_B}\sum_{m=1}^{\infty}\sum_{n=1}^{\infty} p_{mn}\left[\left(\frac{m\pi}{L_1}\right)^2 + \left(\frac{n\pi}{L_2}\right)^2\right]^{-2}\left[1 + \frac{D_B}{D_S}\left[\left(\frac{m\pi}{L_1}\right)^2 + \left(\frac{n\pi}{L_2}\right)^2\right]\right]$$
$$\sin\left(\frac{m\pi}{L_1}X_1\right)\sin\left(\frac{n\pi}{L_2}X_2\right) \qquad \forall m, n \in \mathbb{N}_\star^+ \tag{6.49}$$

Results for the maximum deflection $w_{max} = w(L_1/2, L_2/2)$ are compared in Table 6.5. The error is far below 1%.

If external loads are excluded in Eq. (6.42), the following equation can be derived.

Table 6.5 Resulting maximum deflection of the shear soft solutions in comparison

Closed-form (Mindlin)	Computational (XLWT)	Deviation	
w_{max} (mm)	w_{max} (mm)	$\Delta_{w_{max}}$ (mm)	$\delta_{w_{max}}$ (%)
$13.610 \cdot 10^{-4}$	$13.624 \cdot 10^{-4}$	$1.4 \cdot 10^{-6}$	0.11

$$\left[\nabla^2 - \frac{\rho_{\circledast} H}{D_S}\frac{\partial^2}{\partial t^2}\right]\left[D_B\nabla^2 - \frac{\rho_{\circledast} H^3}{12}\frac{\partial^2}{\partial t^2}\right] w + \rho_{\circledast} H \frac{\partial^2 w}{\partial t^2} = 0 \tag{6.50}$$

If identical constraints as with the Kirchhoff plate are used to solve the problem, the following approach can be used.

$$w(\boldsymbol{r}, t) = W(\boldsymbol{r})\mathrm{e}^{\iota\omega t} = \sum_{m=1}^{\infty}\sum_{n=1}^{\infty} w_{mn} \sin\left(\frac{m\pi}{L_1}X_1\right)\sin\left(\frac{n\pi}{L_2}X_2\right)\mathrm{e}^{\iota\omega t} \tag{6.51}$$

The Kirchhoff solution known from the preceding section should be used as an abbreviation at this point.

$$\omega_{mn}^{\mathrm{K}} = \sqrt{\frac{D_B}{\rho_{\circledast} H}}\left(\frac{m^2}{L_1^2} + \frac{n^2}{L_2^2}\right) \tag{6.52}$$

At this point correction coefficients for rotational inertia KK_{mn}^{R} and transverse shear KK_{mn}^{Q} are introduced.

$$\left(KK_{mn}^{R}\right)^2 = \frac{H^2}{12}\left(\frac{m^2}{L_1^2} + \frac{n^2}{L_2^2}\right) \qquad \left(KK_{mn}^{Q}\right)^2 = \frac{D_B}{D_S}\left(\frac{m^2}{L_1^2} + \frac{n^2}{L_2^2}\right) \tag{6.53}$$

The result is the eigenfrequency equation for the shear soft plate.

$$f_{mn} = f_{mn}^{\mathrm{K}} \frac{\sqrt{1 + (KK_{mn}^{R})^2 + (KK_{mn}^{Q})^2 \pm \sqrt{(1 + (KK_{mn}^{R})^2 + (KK_{mn}^{Q})^2)^2 - (2KK_{mn}^{R}KK_{mn}^{Q})}}}{\sqrt{2}KK_{mn}^{R}KK_{mn}^{Q}} \quad \forall\, m, n \in \mathbb{N}^+ \tag{6.54}$$

Table 6.6 Resulting natural frequencies of the shear soft solutions in comparison

Closed-form (Mindlin)									
mn	11	12	13	21	22	23	31	32	33
f_{mn} (Hz)	0.64	2.16	4.70	1.02	2.54	5.08	1.65	3.17	5.71
Computational (XLWT)									
r	1	2	3	4	5	6	7	8	9
f_r (Hz)	0.63	2.16	4.69	1.02	2.54	5.07	1.65	3.17	5.69
Deviation									
Δ_f (Hz)	−0.01	0.00	−0.01	0.00	0.00	−0.01	0.00	0.00	−0.02
δ_f (%)	−1.56	0.00	−0.21	0.00	0.00	−0.20	0.00	0.00	−0.35

The results for the eigenfrequencies are juxtaposed in Table 6.6. Maximum deviations can be found at the first natural frequency. After that, the computation error is below 0.2%.

6.4 Evaluation

By means of an a priori error estimation, only the global error behavior of the solution can be determined, with which statements about the influence of individual discretization parameters (h^e, *PG*) as well as the convergence speed can be made. However, these statements appear difficult in the context of a verification of numerically generated solutions. Therefore, we first showed a posteriori that the solutions of both problems converge. Since consistency and stability are necessary conditions for the convergence of the solution of mechanical problems, the solution error is thus limited in the entire domain and the discrete problem is transferred to the original differential equations at the limit $h^e_\alpha \rightarrow 0$.

For the closed-form solutions used for comparison, the Navier double-series approach was used. This approach is characterized by its comparatively simple mathematical handling. Both problems, statics and eigenproblem, could be solved with identical, moment-free supports. A detailed description of the illustrated versions can be found in [1]. As alternative approaches single series solutions of Nádai and Levy [5, 6] can be named. However, the solution cost of these alternatives is higher.

In the static structural analyzes, orthogonal, homogeneous loads with an intensity of $q_0 = 10^{-5}\,{}^{\mathrm{N}}\!/_{\mathrm{mm}^2}$ (shear-rigid plate) or $q_0 = 10^{-10}\,{}^{\mathrm{N}}\!/_{\mathrm{mm}^2}$ (shear soft plate) are used. For both scenarios, the terms of the series solutions $m \in \{1 \ldots 9\}$ are sufficient for the Fourier series. The eigenvalue evaluation is limited to the first nine natural frequencies.

The graphical comparisons between closed-form and computational solutions for the shear-rigid and the shear soft case is given in Fig. 6.5. In context of the static analyses, the plate bisecting line was used to compare the results of the deflections. For the comparison of the analyzed natural frequencies the following definition for the correlation of the indices between closed-form and computational solution was defined.

$$r = \begin{cases} \frac{m \cdot n}{m} + (m-1) & \text{if } m = 1 \\ \frac{m \cdot n}{m} + (m+1) & \text{if } m = 2 \\ \frac{m \cdot n}{m} + (m+3) & \text{if } m = 3 \end{cases} \qquad \forall n \in \{1, 2, 3\} \tag{6.55}$$

The conversion into a sorted sequence for the graphical representation of the eigenfrequency values follows the following convention.

$$f_{MN}:\ f_r < f_{(r+1)} \tag{6.56}$$

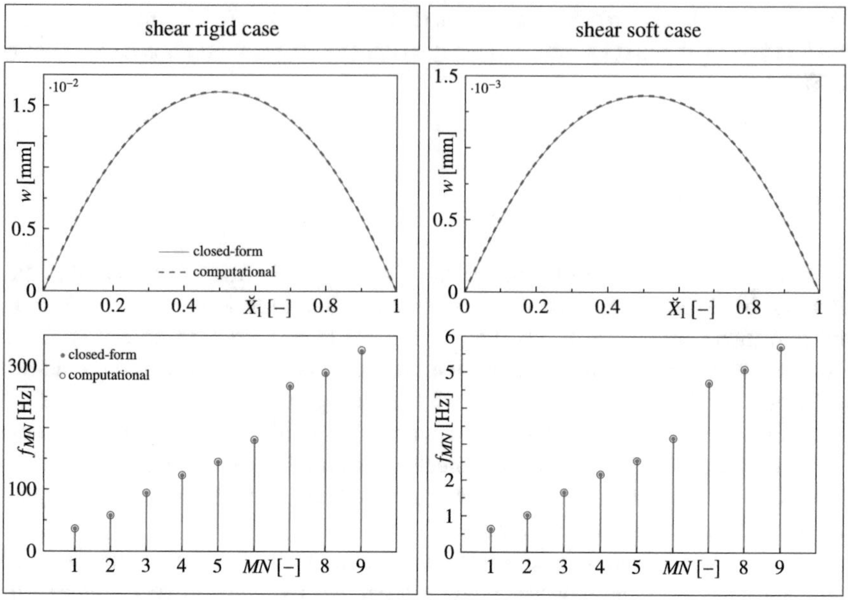

Fig. 6.5 Verification of both monolithic solutions: Shear-rigid and shear-soft case

The computational results show a very good agreement with the closed-form results. The finite element developed is therefore considered verified.

References

1. Altenbach H, Altenbach J, Naumenko K (1998) Ebene Flächentragwerke · Grundlagen der Modellierung und Berechnung von Scheiben und Platten. Springer, Berlin. https://doi.org/10.1007/978-3-642-58721-4
2. Eisenträger J, Naumenko K, Altenbach H, Meenen J (2015) A user-defined finite element for laminated glass panels and photovoltaic modules based on a layer-wise theory. Compos Struct 133:265–277. https://doi.org/10.1016/j.compstruct.2015.07.049
3. Heuser H (1991) Gewöhnliche Differentialgleichungen - Einführung in Lehre und Gebrauch. Teubner, Stuttgart
4. Kirchhoff GR (1850) Über das Gleichgewicht und die Bewegung einer elastischen Scheibe. Journal für die reine und angewandte Mathematik 40:51–88. https://doi.org/10.1515/crll.1850.40.51
5. Levy M (1877) Mémoire sur la théorie des plaques élastiques planes. Journal de Mathématiques Pures et Appliquées 3(3):219–306. http://sites.mathdoc.fr/JMPA/afficher_notice.php?id=JMPA_1877_3_3_A12_0
6. Nádai A (1925) Die elastischen Platten – Die Grundlagen und Verfahren zur Berechnung ihrer Formänderungen und Spannungen, sowie die Anwendungen der Theorie der ebenen zweidimensionalen elastischen Systeme auf praktische Aufgaben. Springer, Berlin. https://doi.org/10.1007/978-3-662-11487-2

7. Navier M (1823) Mémoire sur les lois de l'équilibre et du mouvement des corps solides élastiques. Bulletin de la Société Philomathique de Paris, pp 177–181
8. Timoshenko S, Woinowsky-Krieger S (1987) Theory of Plates and Shells, 2nd edn. McGraw-Hill, New York (1st edn 1959)

Chapter 7
Application

7.1 Preliminary Remarks

At this point, the computational solution strategy introduced in this work is to be applied. The investigations are limited to common data for geometries and materials used. Thereby we focus on data of photovoltaic modules as they are a prominent application of an Anti-Sandwich [1]. A distinction is made between parameter and case studies. In the parameter studies the effects of the variation of geometrical and physical quantities are investigated, while in the case studies examinations are carried out at realistic loading scenarios from natural weathering.

7.2 Parameters Studies

Photovoltaic modules available for terrestrial applications are subject to a large variety of geometrical dimensions and material properties. The durability and reliability of these structures is not insignificantly determined by these structural parameters. In the design process the knowledge of approximately optimal parameters is therefore important. The approach developed here offers a wide range of applications and should be used at this point to make such statements. First, the significant ratios of these parameters as well as the possible variations, as determined in [1], will be recapitulated.

$$TR = \frac{h^c}{h^s} \approx 0.125 \ldots 0.45 \tag{7.1}$$

$$LR = \frac{L_2}{L_1} \approx 0.25 \ldots 1 \tag{7.2}$$

$$TLR = \frac{H}{L_{\min}} \approx 2 \cdot 10^{-3} \ldots 1.4 \cdot 10^{-2} \tag{7.3}$$

M. Aßmus, *Structural Mechanics of Anti-Sandwiches*,
SpringerBriefs in Continuum Mechanics,
https://doi.org/10.1007/978-3-030-04354-4_7

$$GR = \frac{G^c}{G^s} \approx 7 \cdot 10^{-6} \ldots 1.5 \cdot 10^{-2} \tag{7.4}$$

Furthermore, the mass density ratio is introduced, which should be used in the investigation of the natural frequencies.

$$MDR = \frac{\rho^c_{\circledast}}{\rho^s_{\circledast}} \approx 4 \cdot 10^{-2} \ldots 1 \tag{7.5}$$

For subsequent investigations, the geometry and material parameters for core and skin layers introduced here are systematically varied in order to determine the mechanical behavior for the ratios within the limits indicated. For the sake of simplicity we reduce our concern to symmetric composite structures.

$$h^s = h^t = h^b \qquad Y^s = Y^t = Y^b \qquad \nu^s = \nu^t = \nu^b \qquad \rho^s_{\circledast} = \rho^t_{\circledast} = \rho^b_{\circledast}$$

However, the statements regarding the resulting results do not lose their general validity. Starting point are now the structural parameters of Sect. 6.1, which result in the following ratios.

$$\begin{aligned} TR &= 0.15625 \\ LR &= 0.5 \\ TLR &= 9.136 \cdot 10^{-3} \\ GR &= 9.978 \cdot 10^{-5} \\ MDR &= 0.384 \end{aligned} \tag{7.6}$$

The shear correction factor is kept constant at $\kappa^K = 1$ for all subsequent examinations. The discretization is realized with elements of constant element edge length $h^e_\alpha = 10\,\text{mm}$ to obtain convergence with respect to the sought variable. The composite is supported torque-free while only the deflections at the edges remain locked.

7.2.1 Parameter Variation at Static Loading

In order to obtain information about the mechanical behavior with varying structural parameters, the static load study is limited to a simple load case. The load is constant on the acting surface and it acts orthogonal. Based on the parameters listed in Eq. (7.6), geometry and material data are systematically modified to match variations in the structural parameters within the Eqs. (7.1)–(7.4) to analyze given limits. For evaluation, the deflection is used. Due to the load case, the maximum deflection can be used for the ease of evaluation.

$$w_{\max} = w({}^{L_1}\!/_2, {}^{L_2}\!/_2) \tag{7.7}$$

The following dependencies can be formulated with regard to geometry and material data.

$$w_{\max} = \mathcal{F}\left(L_\alpha, h^K, Y^K, \nu^K\right) \qquad \forall\, \alpha \in \{1,2\} \wedge K \in \{t, c, b\} \tag{7.8}$$

This results in eleven parameters that influence the static structural behavior of the three-layered composite [4]. Due to the introduction of the structural parameters, it is sufficient to examine the dependencies of $w_{\max}$ on the following four parameters.

$$w_{\max} = \mathcal{G}\,(TR, LR, TLR, GR) \tag{7.9}$$

For reasons of comparability, the maximum deflection of all variations is normalized such that $0 \leq \overline{w}_{\max} \leq 1$ holds.

$$\overline{w}_{\max}(\Box) = \frac{w_{\max}(\Box)}{\max\left[w_{\max}(\Box)\right]} \qquad \forall\, \Box \in \{TR, LR, TLR, GR\} \tag{7.10}$$

The advantage of such normalization is the achievement of generalized representations and universal statements. The basic goal is to keep the maximum deflections as low as possible, which is made possible by a high structural rigidity. The results of the investigations are illustrated in Fig. 7.1. The dependence of the deflection on the thickness ratio TR is shown in the upper left corner. A parabolic curve can be identified, with low deflections at low TR and large deflections at high TR. Dependencies of the length ratio LR are shown on the top right. Here, too, a parabolic curve with small deflections at low LR and large deflections at high LR is shown. The thickness-to-length ratio is shown at the bottom left. It shows a hyperbolic curve of the deflection dependence. Ratios up to a value of $TLR = 0.5 \cdot 10^{-2}$ appear particularly critical as large maximum deflections occur. For values $TLR > 0.5 \cdot 10^{-2}$ the function is much more flat. At the bottom right of Fig. 7.1 the dependence of the shear modulus GR is shown. Significant areas for applicable theories are color coded. You can see a smooth non-linear curve with large bends at low GR and small bends at high GR. Two borderline cases are identifiable, to which the function approaches each asymptotically. These upper and lower limits can be interpreted as follows.

- monolithic limit
  - loosely dashed line at $\overline{w}_{\max}(GR) \approx 0.15$
  - represents the behavior of a homogeneous single layer
- layerwise limit
  - densely dashed at $\overline{w}_{\max}(GR) = 1$
  - represents the behavior of several individual layers sliding on each other.

For the maximum deflection, the layerwise limit is about 6.6 times higher than the monolithic limit. The monolithic boundary represents the exact solution of the classical laminate theory (CLT), which follows the Kirchhoff theory for shear-rigid plates. Following the course towards decreasing *GR* up to $GR = 10^{-5}$, the first-order shear deformation theory (FOSDT) can be applied [11]. The advantage of the extended layerwise theory (XLWT) is clearly shown here. It is applicable throughout the whole shear modulus ratio range.

7.2.2 Parameter Variation at Eigenfrequency Analysis

The investigations of the eigenbehavior take place on an unloaded structure. The eigenfrequencies are used for the evaluation. The first natural frequency $f_1 = \min(f_r)$ is used as the evaluation criterion. The following dependencies can be specified for this.

$$f_1 = \mathcal{J}\left(L_\alpha, h^K, Y^K, \nu^K, \rho^K_{\circledast}\right) \qquad \forall \alpha \in \{1, 2\} \wedge K \in \{t, c, b\} \tag{7.11}$$

Overall, 14 geometry and material parameters affect the eigenbehavior of the structure [5]. Due to the introduced structural parameters (7.1)–(7.5), it is sufficient to investigate the following dependencies.

$$f_1 = \mathcal{K}\left(TR, LR, TLR, GR, MDR\right) \tag{7.12}$$

For reasons of comparability, the eigenfrequencies of all variations are also normalized at this point, so that $0 \leq \overline{f}_1 \leq 1$ holds.

$$\overline{f}_1(\square) = \frac{f_1(\square)}{\max\left[f_1(\square)\right]} \qquad \forall \square \in \{TR, LR, TLR, GR, MDR\} \tag{7.13}$$

The excitation spectrum for outdoor weathering and transport of photovoltaic modules is given in the literature as up to 50 Hz [2, 12]. Therefore, it is advantageous if the natural frequencies of such structures are above this threshold in order not to resonate. The goal of the investigations presented here are thus the highest possible natural frequencies.

The eigenfrequency behavior results are visualized in Fig. 7.2. The dependence on the thickness ratio *TR* is shown in the upper left corner. It shows a hyperbolic dependence, with the maximum at low *TR*, while the minimum can be found at $TR = 0.5$. The results for the length ratio are shown at the top center. The resulting function has a history analogous to *TR* with a minimum at high *LR*. Directly to the right is the dependency of the thickness-to-length ratio. The course of the resulting function is nearly linear, slightly parabolic. The minimum is below $TLR = 2 \cdot 10^{-3}$ while the maximum is above $TLR = 1.4 \cdot 10^{-2}$. The mass density ratio is shown at the bottom left. The dependence of the first natural frequency on this ratio is completely linear with a minimum at $MDR = 1$ and a maximum at $MDR = 0$. Results on the behavior of the shear modulus ratio are shown in the lower right corner. As with the deflection behavior, two limits can be identified. The interpretation of these limits is

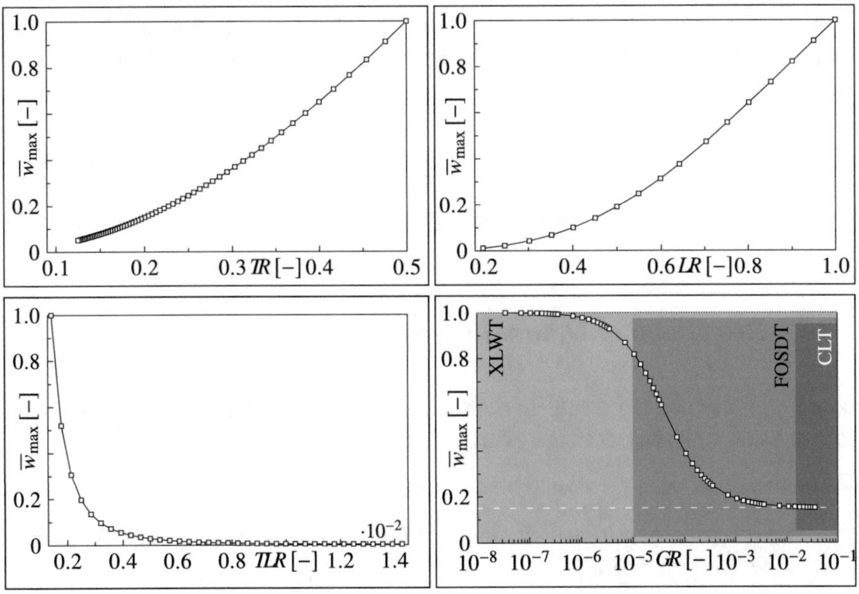

Fig. 7.1 Deflection behavior with geometry and material variation of the structure

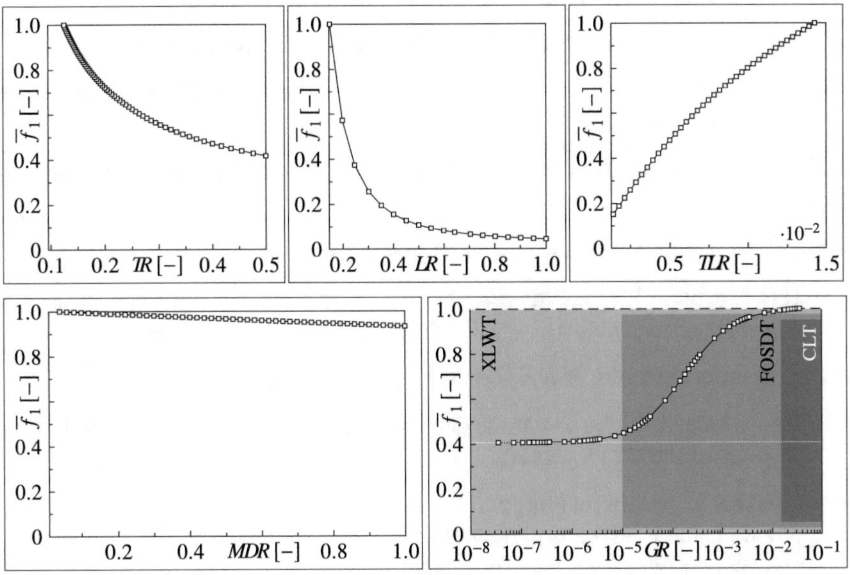

Fig. 7.2 Eigenfrequency behavior with geometry and material variation of the structure

identical to that of Sect. 7.2.1. It is striking, however, that the function is mirrored vertically. The lower bound (dashed line at $\overline{f}_1 \approx 0.4$) is now the layerwise limit and the upper bound (loosely dashed at $\overline{f}_1 = 1$) represents the monolithic limit. The monolithic limit is about 2.5 times higher than the layered limit. The areas of application of the available theories are also marked in color. Here, too, the broad scope of the XLWT is apparent. Finally, it should be noted that the variation of the mass density ratio has the least influence on the results compared to all other ratios.

7.2.3 Discussion of Optimal Parameters

Based on the generated results in both preceding sections, the following statements can be made with respect to the geometric relationships.

- The thickness ratio *TR* should be as small as possible.
  - This is achieved by a small thickness of the core layer h^c and/or large thicknesses of the skin layers h^s.
- The length ratio *LR* should be as small as possible.
  - This is achieved by the most divergent plane dimensions L_α.
- The thickness-to-length ratio *TLR* should be as large as possible.
  - This is achieved by the largest possible total thickness H and/or small minimum plane dimension $L_{\min}$.

With regard to the ratios of material parameters, the following statements may be implied.

- The shear modulus ratio *GR* should be as large as possible.
  - This is achieved by the highest possible shear modulus of the core layer G^c. The order of magnitude $G^c \leq 10^{-3} G^s$ should be the target value.
- The mass density ratio MDR should be as small as possible.
  - This is achieved by a low mass density of the core layer $\rho^c_{\circledast}$ compared to the densities of the skin layers $\rho^s_{\circledast}$.

Finally, it should be noted at this point that the statements made here in the area of the design and planning phase of photovoltaic modules for dimensioning and material selection are considered particularly valuable.

7.3 Case Studies

7.3.1 Parametrization

As mentioned, photovoltaic modules are a widely used and popular application for Anti-Sandwiches [1]. Nowaday, these components are omnipresent. They are installed on rooftops or solar parks. Thereby they are exposed to natural weathering during their whole service life. Here, wind and snow loading are the most significant loading scenarios. In this section, we will investigate the effects on the structural behavior using two exemplary loading scenarios.

In the context of the case studies, loads are only supposed to act on the front side $\mathfrak{V}$ of the Anti-Sandwich. The load vector is composed as follows.

$$q\left(r\right) = s\left(r\right) + p\left(r\right) \qquad \forall\, r \in \mathfrak{V} \tag{7.14}$$

Herein $s = -s_\alpha e_\alpha$ are in-plane loads and $p = pn$ are loads orthogonal to the plane. We still have $n \equiv e_3$ and $r = X_\alpha e_\alpha$. In order to be able to carry out realistic investigations, the photovoltaic module should be inclined at the angle θ. From Fig. 7.3 it can be seen that in both load cases a distinction has to be made on the basis of the direction of acting loads. First, the consideration of the angle of attack will be presented. For this purpose, a fixed coordinate system g_i is introduced next to the composite coordinate system. Since the employment of the photovoltaic-module is limited here to the consideration of the height of the sun, it is sufficient to rotate around the axis $e_2 \equiv g_2$ in order to be able to move from one coordinate system to another transform.

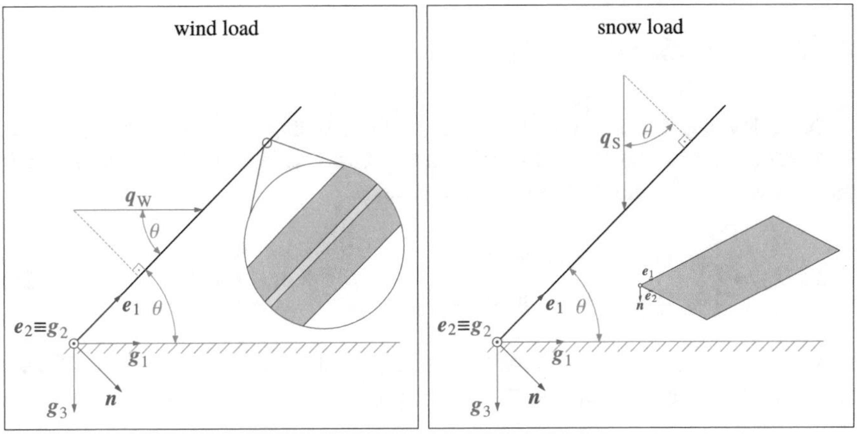

Fig. 7.3 Tilt angle θ as well as coordinate systems $\{g_i\}$ and $\{e_\alpha, n\}$ used for parametrization

$$\boldsymbol{g}_1 = \cos(\theta)\,\boldsymbol{e}_1 + \sin(\theta)\,\boldsymbol{e}_3 \tag{7.15}$$

$$\boldsymbol{g}_2 = \boldsymbol{e}_2 \tag{7.16}$$

$$\boldsymbol{g}_3 = -\sin(\theta)\,\boldsymbol{e}_1 + \cos(\theta)\,\boldsymbol{e}_3 \tag{7.17}$$

Conversely, the following correlations apply.

$$\boldsymbol{e}_1 = \cos(\theta)\,\boldsymbol{g}_1 - \sin(\theta)\,\boldsymbol{g}_3 \tag{7.18}$$

$$\boldsymbol{e}_2 = \boldsymbol{g}_2 \tag{7.19}$$

$$\boldsymbol{e}_3 = \sin(\theta)\,\boldsymbol{g}_1 + \cos(\theta)\,\boldsymbol{g}_3 \tag{7.20}$$

The transition can be realized with an orthogonal tensor $\boldsymbol{Q}$.

$$\boldsymbol{Q} = \boldsymbol{e}_i \otimes \boldsymbol{g}_i \tag{7.21}$$

Considering the Eqs. (7.15)–(7.20), we obtain the following representation for this versor and its coefficient matrix with respect to an orthogonal basis.

$$\boldsymbol{Q} = Q_{ij}\,\boldsymbol{e}_i \otimes \boldsymbol{e}_j = Q_{ij}\boldsymbol{g}_i \otimes \boldsymbol{g}_j \qquad \left[Q_{ij}\right] = \begin{bmatrix} \cos(\theta) & 0 & \sin(\theta) \\ 0 & 1 & 0 \\ -\sin(\theta) & 0 & \cos(\theta) \end{bmatrix} \tag{7.22}$$

It is assumed that a load vector $\boldsymbol{a} = a_i \boldsymbol{g}_i$ exists. The components a_i' in the rotated coordinate system $\{\boldsymbol{e}_i\}$ are then calculated as follows.

$$a_i' = Q_{ji} a_j \tag{7.23}$$

7.3.2 Approximation of Loading

In the following, the equations from the previous section are used to approximate loads resulting from the natural weathering spectrum. It is assumed that wind loads $\boldsymbol{q}_{\mathrm{W}}$ act horizontally and snow loads $\boldsymbol{q}_{\mathrm{S}}$ vertically.

$$\boldsymbol{q}_m(\boldsymbol{r}) = \begin{cases} q_{\mathrm{W}}(\boldsymbol{r})\,\boldsymbol{g}_1 & \text{wind load} \\ q_{\mathrm{S}}(\boldsymbol{r})\,\boldsymbol{g}_3 & \text{snow load} \end{cases} \qquad \forall m \in \{\mathrm{W}, \mathrm{S}\} \tag{7.24}$$

Herein $\boldsymbol{q}_m$ are amplitude functions that need to be defined. Utilizing Eq. (7.22) we get the following representation for the composite coordinate system.

$$\boldsymbol{q}_m(\boldsymbol{r}) = \begin{cases} q_{\mathrm{W}}(\boldsymbol{r})\cos(\theta)\,\boldsymbol{e}_1 + q_{\mathrm{W}}(\boldsymbol{r})\sin(\theta)\,\boldsymbol{n} & \text{wind load} \\ -\,q_{\mathrm{S}}(\boldsymbol{r})\sin(\theta)\,\boldsymbol{e}_1 + q_{\mathrm{S}}(\boldsymbol{r})\cos(\theta)\,\boldsymbol{n} & \text{snow load} \end{cases} \tag{7.25}$$

To investigate the influence of locally heterogeneous loads, the load distribution in the coordinate system $\boldsymbol{g}_i$ is defined as in Eq. (7.24). Basically, arbitrary amplitude functions $q_m(\boldsymbol{r})$ can be taken into account. As an example, the double sine function applies.

$$q_m(\boldsymbol{r}) = q_m^0 \sin\left(\frac{\pi}{L_1}X_1\right)\sin\left(\frac{\pi}{L_2}X_2\right) \tag{7.26}$$

Here, $q_m^0 \;\forall\, m \in \{\mathrm{W}, \mathrm{S}\}$ is the maximum amplitude. The representation from Eq. (7.26) can be generalized by means of a Fourier series, whereby e.g. experimentally determined load distributions can be approximated.

$$q_m(\boldsymbol{r}) = \sum_{i=1}^{k}\sum_{j=1}^{k} \left(q_m^0\right)_{ij} \sin\left(\frac{\pi i}{L_1}X_1\right)\sin\left(\frac{\pi j}{L_2}X_2\right) \quad \forall m \in \{\mathrm{W}, \mathrm{S}\} \;\wedge\; k \in \mathbb{N}^+ \tag{7.27}$$

Due to the lack of experimental data concerning the distribution functions, a restriction to a Fourier series with only one series term ($k = 1$) takes place here. The load constructed in Eq. (7.26) with respect to the fixed coordinate system $\{\boldsymbol{g}_i\}$ from Eq. (7.24) now has to be rotated to the coordinate system $\{\boldsymbol{e}_i\}$. If Eq. (7.25) is taken into account, the load approximations for wind and snow load result in the following representations.

$$\boldsymbol{q}_{\mathrm{W}}(\boldsymbol{r}) = -s_1\boldsymbol{e}_1 + p\boldsymbol{n} \quad \text{with} \begin{cases} s_1(\boldsymbol{r}) = -q_{\mathrm{W}}^0 \cos(\theta)\sin\left(\frac{\pi}{L_1}X_1\right)\sin\left(\frac{\pi}{L_2}X_2\right) \\ p(\boldsymbol{r}) = q_{\mathrm{W}}^0 \sin(\theta)\sin\left(\frac{\pi}{L_1}X_1\right)\sin\left(\frac{\pi}{L_2}X_2\right) \end{cases} \tag{7.28}$$

$$\boldsymbol{q}_{\mathrm{S}}(\boldsymbol{r}) = -s_1\boldsymbol{e}_1 + p\boldsymbol{n} \quad \text{with} \begin{cases} s_1(\boldsymbol{r}) = q_{\mathrm{S}}^0 \sin(\theta)\sin\left(\frac{\pi}{L_1}X_1\right)\sin\left(\frac{\pi}{L_2}X_2\right) \\ p(\boldsymbol{r}) = q_{\mathrm{S}}^0 \cos(\theta)\sin\left(\frac{\pi}{L_1}X_1\right)\sin\left(\frac{\pi}{L_2}X_2\right) \end{cases} \tag{7.29}$$

These two load vectors are now used to approximate the loads using Eq. (5.47) in the context of the numerical solution strategy.

Table 7.1 Geometry and material data of the composite structure investigated in the case studies

K	L_1^K (mm)	L_2^K (mm)	h^K (mm)	Y^K ($^{N}/_{mm^2}$)	ν^K (–)
t/b	1620	810	3.2	$73.0 \cdot 10^3$ [7]	0.30 [7]
c			1.0	see Table 7.2	

7.3.3 *Dimensions, Temperature-Dependent Material Properties and Boundary Conditions*

For the case studies, the geometry of a commercial photovoltaic module with symmetrical glass-glass construction is used. The core layer should consist of the material ethylene-vinyl acetate (EVA) [9]. Lengths and thicknesses of this structure are exemplary, since there is a wide variety of possible dimensions on the market [1]. The dimensions used here are summarized in Table 7.1. The geometry ratios are identical to those of Eq. (7.6).

$$TR = 0.15625 \qquad LR = 0.5 \qquad TLR = 9.136 \cdot 10^{-3} \tag{7.30}$$

The material parameters for the two skin layers are also given in the Table 7.1.

Deviating from the previously used material characteristics for the soft core layer, the material properties adapted to the load situations should be used in the case studies. This is due to the fact that the most commonly used material is highly temperature-sensitive in terms of Young's modulus. These temperature-dependent mechanical properties are listed in Table 7.2. Characteristic values for $\vartheta = -40°C$ apply to the snow load and characteristic values for $\vartheta = +80°C$ are to apply to the wind load. Furthermore, it is assumed that the material parameters of the skin layers are invariant with respect to temperature changes, at least in the investigated temperature range $\vartheta = -40 \cdots + 80°C$. The temperature-dependent shear modulus GR, which deviates from previous investigations, can also be found in the Table 7.2.

For the studies presented so far, identical boundary conditions were used on all four margins, since only a reclining composite was investigated. Du to the introduction of the angle of attack $\theta \neq 0$ displacement preventions corresponding to the load cases in order to prevent rigid body movements must be taken into account. This concerns the edge Γ_1^1 for the investigations of the snow load and the margin Γ_1^2 for the investigations of the wind load. The definition of these margins as well as the boundary conditions introduced there can be taken from Table 7.3.

Table 7.2 Temperature-dependent material parameters of the core layer material [9]

ϑ (°C)	Y^c ($^{N}/_{mm^2}$)	ν^c (–)	G^c ($^{N}/_{mm^2}$)	GR (–)
−40	1019.04	0.41	361.36	$1.29 \cdot 10^{-2}$
+80	0.52	0.41	0.18	$6.57 \cdot 10^{-6}$

Table 7.3 Definition of boundaries and constraints at the layerwise structure

Boundary	Γ_1^1	Γ_1^2	Γ_2^1	Γ_2^2	Γ_N
$\boldsymbol{e}_1$	0	L_1	$0\ldots L_1$	$0\ldots L_1$	$0\ldots L_1$
$\boldsymbol{e}_2$	$0\ldots L_2$	$0\ldots L_2$	0	L_2	$0\ldots L_2$
$\boldsymbol{n}$	$-H/2\ldots H/2$	$-H/2\ldots H/2$	$-H/2\ldots H/2$	$-H/2\ldots H/2$	$-H/2$
Wind load	$w=0$	$w=0$ $v_1^K=0$	$w=0$	$w=0$	$q_\mathrm{W}^0=5\cdot 10^{-4}\,\mathrm{N/mm^2}$
Snow load	$w=0$ $v_1^K=0$	$w=0$	$w=0$	$w=0$	$q_\mathrm{S}^0=5\cdot 10^{-4}\,\mathrm{N/mm^2}$

In the following, the evaluations of the investigations are presented separately. The results are reduced to representations of paths of the respective plate bisectors. This is on the one hand the path A-A ($X_1=0\ldots L_1$, $X_2=L_2/2$) and on the other hand the path B-B ($X_1=L_1/2$, $X_2=0\ldots L_2$). Besides a normalization of the coordinates in the form $\breve{X}_\alpha=X_\alpha/L_\alpha$ is used. The individual degrees of freedom as well as the kinematic and kinetic quantities are used for the evaluation. The degrees of freedom are presented under consideration of different angles of attack θ. With regard to the resulting kinematic and kinetic quantities, the result representation is reduced to an angle of attack of $\theta=35°$ for reasons of clarity. This value represents a typical mean value for photovoltaic modules installed in Germany (approx. latitude $47°\ldots52°$ N [10]), since the optimal position is always orthogonal to the solar radiation. Due to the symmetry of the tensors $\boldsymbol{N}^K$, $\boldsymbol{L}^K$, $\boldsymbol{G}^K$ and $\boldsymbol{K}^K$ the following representation is limited to the elements $\square_{11}^K$, $\square_{22}^K$ and $\square_{12}^K$ $\forall\square\in\{N,M,G,K\}$.

7.3.4 Wind Loading

Figure 7.4 shows the resulting degrees of freedom of the wind load investigations at elevated temperatures for the paths A-A (X_1) and B-B (X_2). It is shown that the deflection increases with increasing angle of attack. In contrast to purely orthogonal loading [3, 8], the deflection curve is asymmetrical. The maximum deflection is not in the plate center ($w_\mathrm{max}\neq w(L_1/2, L_2/2)$). The maximum value is $w_\mathrm{max}=1.95\,\mathrm{mm}$ for an angle of attack of $\theta=55°$. The in-plane displacements reach their extreme values at $\breve{X}_1=0$, where $v_1^K(\breve{X}_1=1)=0$ applies due to the introduced boundary condition. The absolute values of the displacements $|v_2^K|$ are identical for front and back layer along B-B, but in opposite directions. The shifts of the core layer are nearly 0. In contrast, the rotations φ_1^K and φ_2^K of the skin layers are identical, while the rotation of the core layer is opposite.

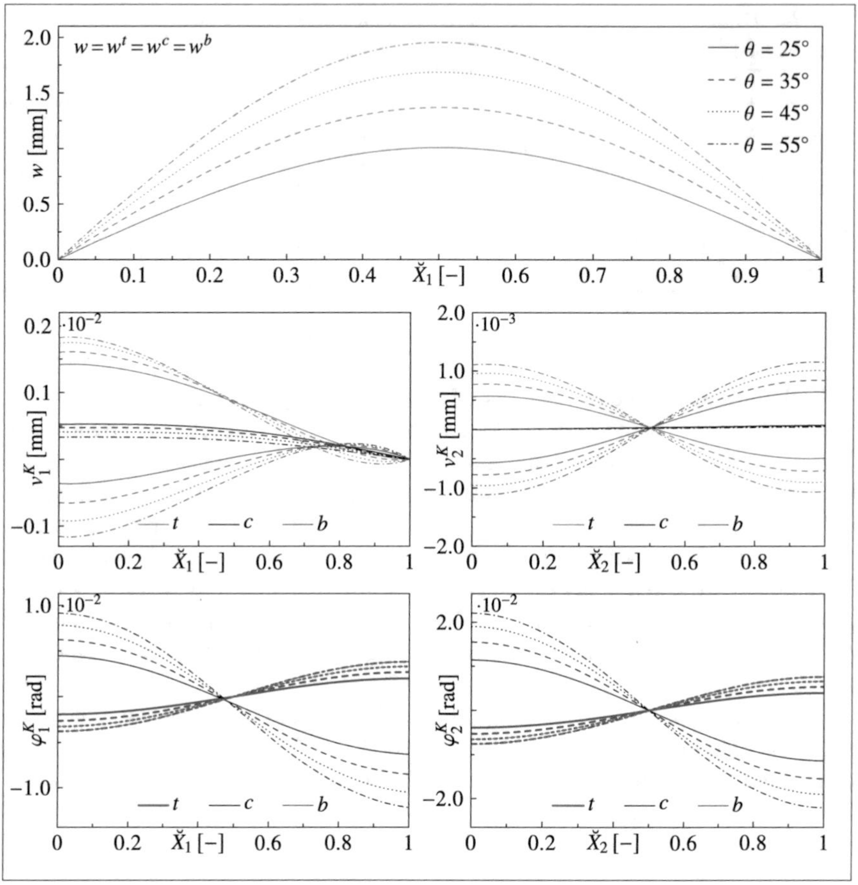

Fig. 7.4 Deflections, displacements and rotations at wind loading

Figure 7.5 shows the results of the stress resultant (both upper rows) and the conjugated kinematic quantities (both lower rows) for the path A-A. For the membrane forces, $N_{11}^K(\check{X} = 0) = 0$ is valid, since the boundary Γ_1^1 is free of displacement inhibitions. Furthermore the membrane shear terms N_{12}^K are about 0. There are no membrane forces in the core layer. Since deflections are blocked at the edges, typical boundary layer effects occur at the transverse shear forces [6]. A mesh refinement in the boundary area would reduce this effect. The bending moments $M_{\alpha\alpha}^K$ are 0 at the edges $\check{X}_1 = 0$ and $\check{X}_1 = 0$, because the rotations along the edges Γ_1^1 and Γ_1^2 can form freely. Analogous to the membrane forces N_{12}^K the torsion moments are $M_{12}^K = 0$.

Figure 7.6 shows the results of the corresponding values along the path B-B. The membrane forces in the skin layers reach their extreme value in the center of the plate ($\check{X}_2 = 0.5$). On the other hand, the membrane forces of the core layer are 0. The transverse shear forces Q_1^K are 0 in the middle of the plate and reach their extreme value at the plate edges. The boundary layer effect is also recognizable here for

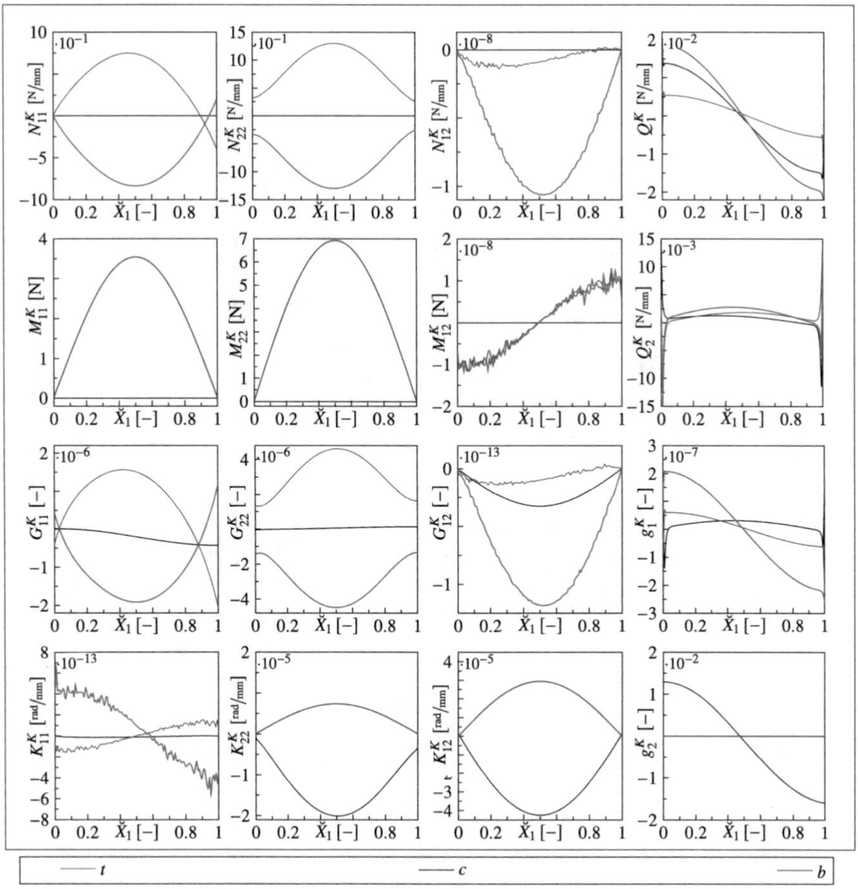

Fig. 7.5 Layer-by-layer kinetic and kinematic measures at wind loading for path A-A

Q_2^K. The bending moments are $M_{\alpha\alpha}^K(\check{X}_1 = 0) = M_{\alpha\alpha}^K(\check{X}_1 = 1) = 0$. The torsional moment M_{12}^K is influenced by a boundary layer effect. Here, too, the core layer shows itself free of moments. The transversal shear strains are negligible.

7.3.5 *Snow Loading*

Since snow loads usually occur at temperatures below freezing point, the material characteristic of the core layer is used for $\vartheta = -40°\text{C}$. Figure 7.7 shows the resulting degrees of freedom of the snow load investigations at reduced temperatures. In contrast to the wind load investigations, the deflection here increases with decreasing angle of attack. The maximum occurs at $\theta = 25°$ with a value of $w_{\max} = 0.45$ mm.

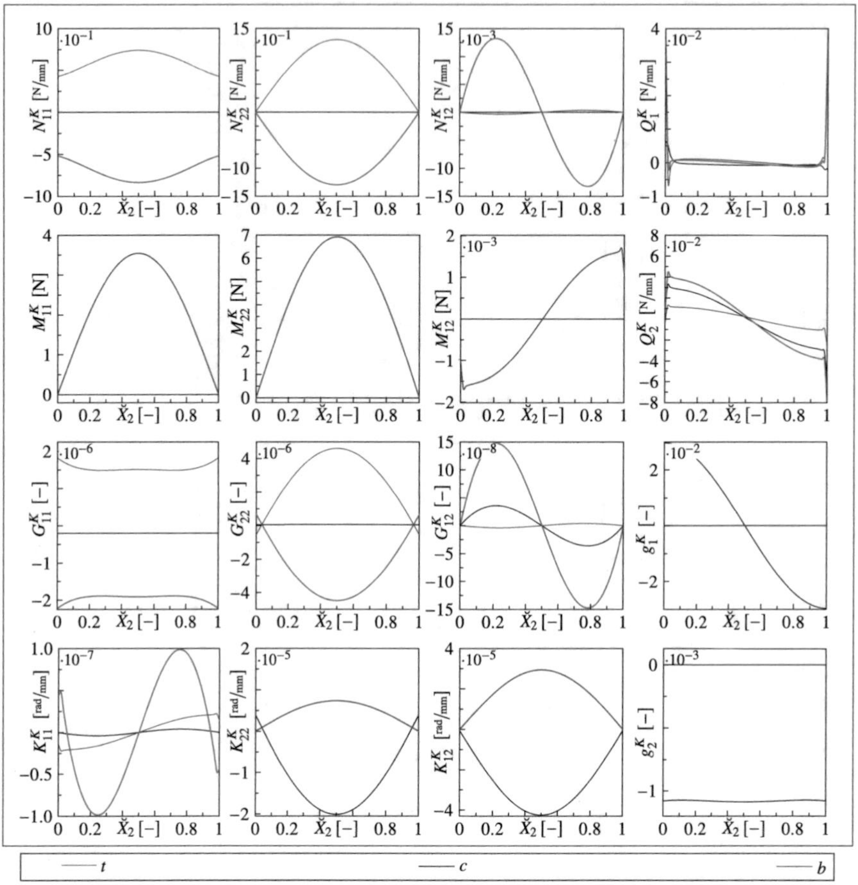

Fig. 7.6 Layer-by-layer kinetic and kinematic measures at wind loading for path B-B

At $\breve{X} = 0$ the deflection change is nearly 0. This is caused by the limited displacements v_1^K. As a result, the mutual slip of the two skin layers is significantly reduced at reduced temperatures due to the increased shear stiffness of the core layer. Therefore also the rotations φ_1^K disappear at this point. Even under snow load, the maximum deflection is outside the center of the plate. The displacements v_1^K reach their maximum at $\breve{X} = 1$. Due to the symmetry of the boundary conditions with respect to $\breve{X}_2$ the displacements v_2^K reach their highest values at the edges and are 0 in the center of the module. The displacements of the top layer is contrary to those of the back layer. The displacements of the core layer $v_2^c(\breve{X}_2)$ are 0. Due to the increased shear stiffness of the core layer, the rotations of all layers are equally directed. The rotations of the cover layers are identical while the core layer is slightly smaller. The rotations φ_2^K are 0 at $\breve{X}_2 = 0.5$ and reach their maximum or minimum at the edges.

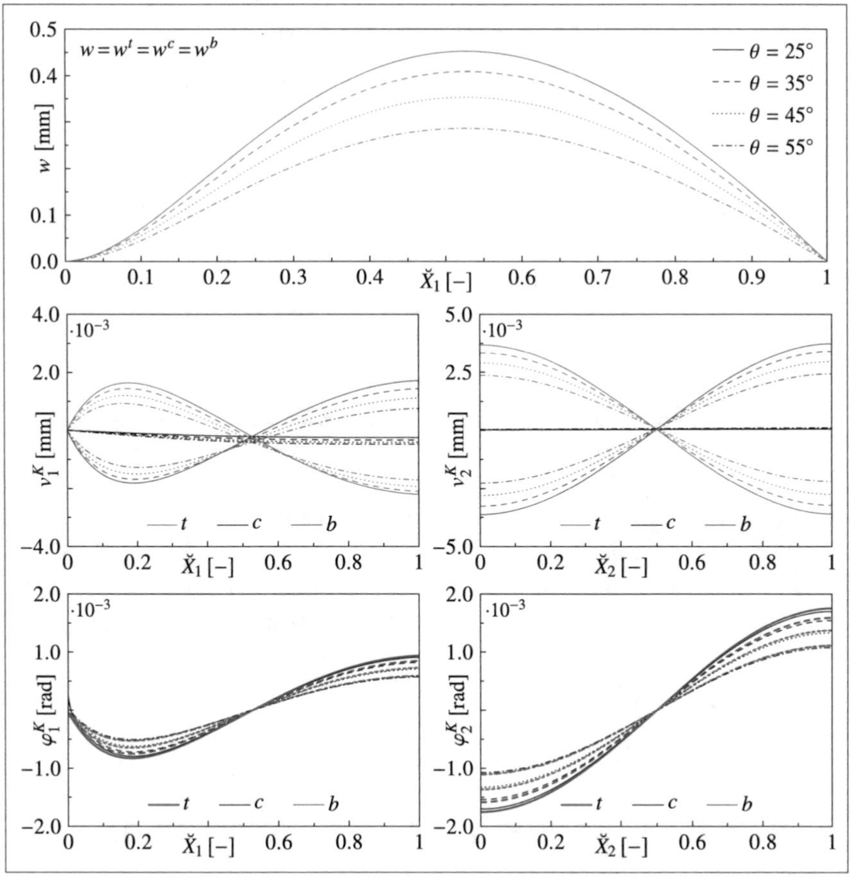

Fig. 7.7 Deflections, displacements and rotations at snow loading

Figure 7.8 shows the results of the stress resultants (both upper rows) and the conjugated kinematic quantities (both lower rows) for path A-A at snow loading. Due to the different boundary conditions for the displacements in comparison to wind loading, the membrane forces show a different behavior. In analogy to wind loading, however, the membrane shear forces are approximately 0. Also here boundary layer effects can be found for the transverse shear terms Q_α^K. The behavior of the bending moments $M_{\alpha\alpha}^K$ deviates strongly in the range $\check{X}_1 = 0$ from the behavior at wind loading. However, the core layer is free of moments and the torsion moments are 0 also. The membrane strains G_{11}^K reach their maximum at $\check{X}_1 = 0$ and become 0 at the opposite edge. Altogether the membrane strains show a similar behavior as

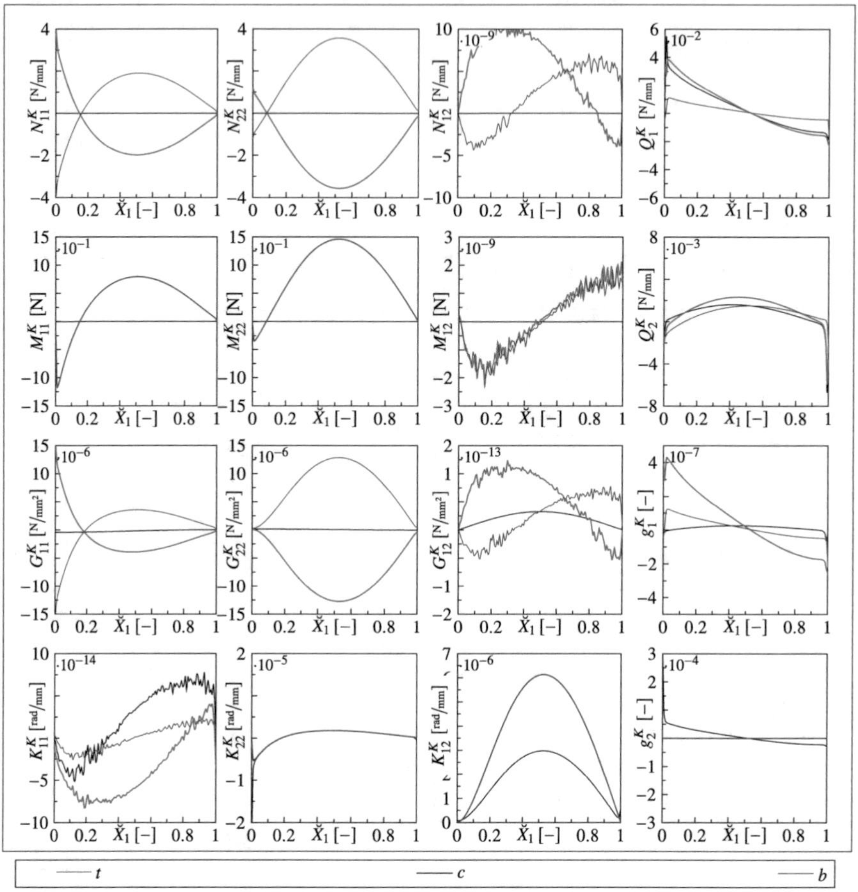

Fig. 7.8 Layer-by-layer kinetic and kinematic measures at snow loading for path A-A

the corresponding membrane forces. Again, the membrane shear strains are nearly 0 and the transverse shear strain g_2^K disappears. Also the boundary layer effects of the transversal shear strains are present. The curvature changes K_{11}^K are 0 in all layers. The curvature change of the core layer K_{22}^c reaches a minimum at $\check{X}_1 = 0$ and tends towards 0 at $\check{X}_1 = 1$.

Figure 7.9 shows the results of the corresponding quantities along the path B-B. Basically, the membrane strains have a similar course as with wind loads, which are shown in Fig. 7.6. However, these are significantly higher with snow loading. The membrane shear strains are several orders of magnitude higher at wind loading due to the higher temperatures compared to snow loading at low temperatures.

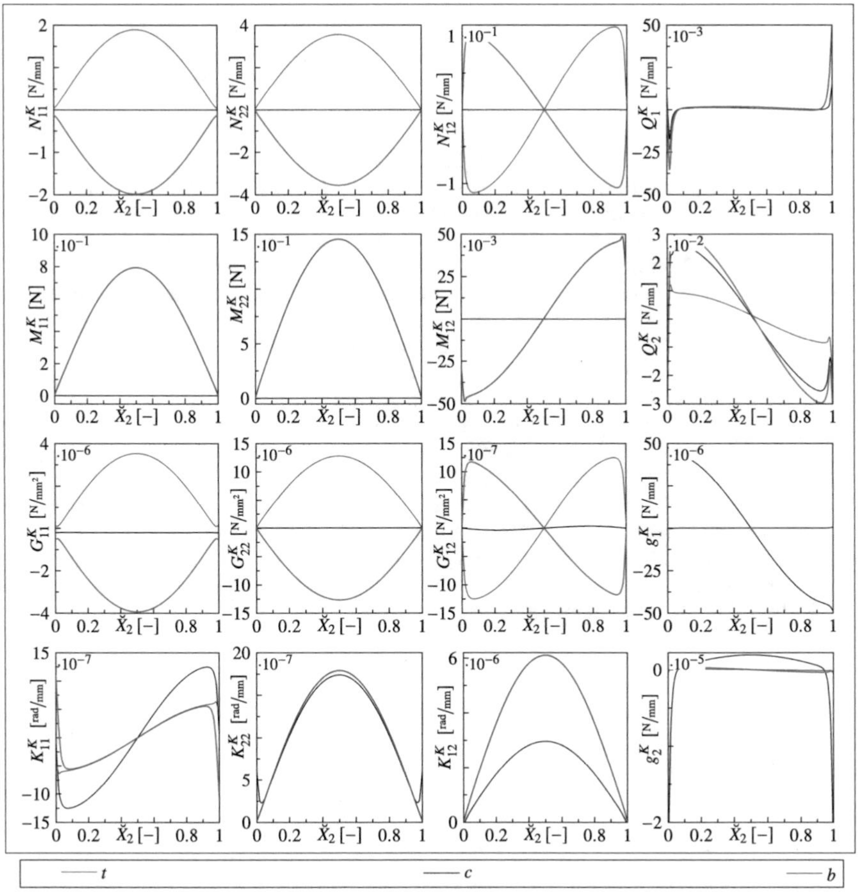

Fig. 7.9 Layer-by-layer kinetic and kinematic measures at snow loading for path B-B

References

1. Aßmus M (2018) Global structural analysis at photovoltaic modules: theory, numerics, application (in German). Dissertation, Otto von Guericke University Magdeburg
2. Aßmus M, Köhl M (2012) Experimental investigation of the mechanical behavior of photovoltaic modules at defined inflow conditions. J Photonics Energy 2(1):1–11. https://doi.org/10.1117/1.JPE.2.022002
3. Aßmus M, Naumenko K, Altenbach H (2016) A multiscale projection approach for the coupled global-local structural analysis of photovoltaic modules. Compos Struct 158:340–358. https://doi.org/10.1016/j.compstruct.2016.09.036
4. Aßmus M, Bergmann S, Naumenko K, Altenbach H (2017) Mechanical behaviour of photovoltaic composite structures: a parameter study on the influence of geometric dimensions and material properties under static loading. Compos Commun 5(-):23–26. https://doi.org/10.1016/j.coco.2017.06.003

5. Aßmus M, Naumenko K, Altenbach H (2017) Mechanical behaviour of photovoltaic composite structures: influence of geometric dimensions and material properties on the eigenfrequencies of mechanical vibrations. Compos Commun 6(-):59–62. https://doi.org/10.1016/j.coco.2017.10.003
6. Brank B (2008) On boundary layer in the Mindlin plate model: Levy plates. Thin-Walled Struct 46(5):451–465. https://doi.org/10.1016/j.tws.2007.11.003
7. Brückner R (2006) Mechanical properties of glasses. In: Cahn RW, Haasen P, Kramer EJ, Zarzycki J (eds) Glasses and amorphous materials, materials science and technology: a comprehensive treatment, vol 9. Wiley-VCH, Weinheim, pp 665–713. https://doi.org/10.1002/9783527603978.mst0102
8. Eisenträger J, Naumenko K, Altenbach H, Meenen J (2015) A user-defined finite element for laminated glass panels and photovoltaic modules based on a layer-wise theory. Compos Struct 133:265–277. https://doi.org/10.1016/j.compstruct.2015.07.049
9. Eitner U (2011) Thermomechanics of photovoltaic modules. Dissertation, Martin-Luther-Universitat Halle-Wittenberg. urn:nbn:de:gbv:3:4-5812
10. Huld T, Šúri M, Dunlop ED (2008) Comparison of potential solar electricity output from fixed-inclined and two-axis tracking photovoltaic modules in europe. Prog Photovolt Res Appl 16(1):47–59. https://doi.org/10.1002/pip.773
11. Naumenko K, Eremeyev VA (2014) A layer-wise theory for laminated glass and photovoltaic panels. Compos Struct 112:283–291. https://doi.org/10.1016/j.compstruct.2014.02.009
12. Sander M, Ebert M (2009) Vibration analysis of pv modules by laser-doppler vibrometry. In: 24th European photovoltaic solar energy conference (PVSEC), pp 3446–3450. https://doi.org/10.4229/24thEUPVSEC2009-4AV.3.46

Chapter 8
Summary and Outlook

8.1 Summary and Assessment

In present treatise an approach for structural analysis of Anti-Sandwiches is presented. In principle, theories for thin-walled structures are suitable for mechanical analysis at such configurations. Since mechanical properties and structural thicknesses of the different layers of an Anti-Sandwich differ widely, classical approaches for composite structures fail to predict correct results. Therefore, a layer-wise approach is chosen within the present discourse. Each layer is considered as a single continuum, while all equations are related to the middle surface of the structure.

First, the direct approach is used to set up the underlying theory for two dimensional continua. In addition to the displacement field, this continuum also possesses a rotation field for describing the kinematics. This concept has physical weaknesses, but results in a mathematically well-posed problem for the efficient treatment of boundary value problems. The description is limited to geometric and physical linearity. The work follows the direct notation. This allows a concise presentation of the basic equations, but without completely falling into the Bourbakian style [7]. Furthermore, a formalization of the governing equations of the plane two dimensional continuum is carried out.

Based on this, the coupling of three layers of such a continuum to a composite is presented. In the process, assumptions and restrictions are discussed. The result is the formulation of the boundary value problem of the overall structure for a three-layered composite structure. The strategy is now known as extended layer-wise theory which is a generalized approach for three-layered composite structures with arbitrary mechanical stiffnesses and structural thicknesses of the layers.

Since the solution of the formulated boundary value problem in closed form usually tightens a too narrow a frame for practical problems, a procedure for the numerical treatment by means of the finite element method is introduced. Therefore, a variational principle is exploited to gain a weak form of governing equations. This form is used to derive the discretized equation of motion. By using a classic

M. Aßmus, *Structural Mechanics of Anti-Sandwiches*,
SpringerBriefs in Continuum Mechanics,
https://doi.org/10.1007/978-3-030-04354-4_8

finite element type and through the consideration of artificial stiffening effects, the numerical formulation gains in efficiency and accuracy. Although the extension of the element by several kinematic degrees of freedom is laborious, it follows the classical systematology.

Numerical examples demonstrate the reliability of the developed finite element solution strategy. Two closed-form comparative examples serve to check the solutions generated numerically. Here, limiting cases of mechanical behavior are used. Parameter and case studies complete the application where important conclusions on the effects of geometry and material data are drawn.

The strategy developed here is particularly useful in the design and development phase of Anti-Sandwiches. With the numerical solution approach provided here, it is possible to predict the global structural behavior as early as in the product development process, which can save high costs for experimental analyzes.

8.2 Conclusion and Outlook

All achievements presented in present work are based on the assumption of geometric and physical linearity. In order to build a generalized theory, these restrictions need be relaxed. The assumption of linear elasticity is based on the strongest restriction on the influence of past deformations on actual stresses. Especially with regard to the physical effect of material nonlinearity, this situation does not seem unproblematic. First of all, the question arises as to how non-linear-elastic material behavior can be mapped in the sense of a St. Venant–Kirchhoff, Neo–Hooke or Mooney–Rivlin material model. In addition, the shear-soft core layer often exhibits viscoelastic material behavior. Altenbach [1] shows an extension by incorporating a viscoelastic material law. This procedure should be considered useful for mapping the relaxation and retardation behavior in the core layer. In this context, however, at least large rotations at the core layer must be taken into account, which result from the indicated behavior of the mutual sliding of the cover layers. In addition to the large deflections, large rotations thus play a decisive role. The Föppl–von Kármán equations often used for moderate deformations [8, 16] can not be used. These equations and their basic assumptions provoke some discussion in the framework of continuum mechanics [2]. It remains unclear which placement they refer to. A consequent extension on the basis of the theory presented in Chap. 2 thus concerns the geometrically nonlinear formulation for the consideration of finite deflections, rotations and displacements. It is necessary to check to what extent the restrictions introduced for the coupling of the layers to the composite can be maintained.

Another aspect of developments concerns the time-dependent behavior. So far, only static influences have been investigated, and the influence of inertia for eigenvalue analysis has been considered. As shown in [3], the load spectrum also consists of dynamic components. In addition, studies on the absorption behavior of Anti-Sandwiches are required to fit the damping matrix into the structural Eq. (5.57) and to predict the dynamic behavior in case of external excitation. In addition to the

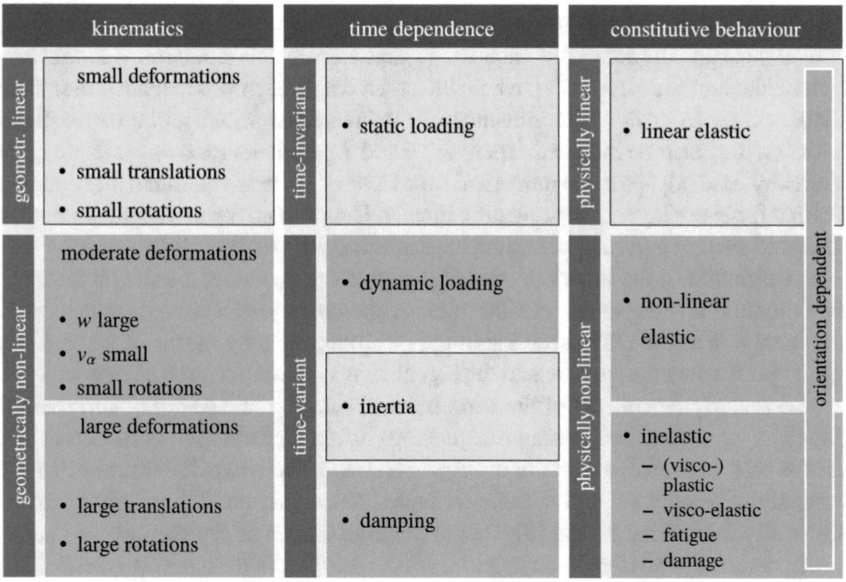

Fig. 8.1 Edited and open fields of work for global structural analysis of Anti-Sandwiches

spatial discretization, a temporal discretization has to be realized, whereby explicit and implicit methods for solving the structural equation $\mathbf{M}\ddot{\mathbf{a}} + \mathbf{D}\dot{\mathbf{a}} + \mathbf{K}\mathbf{a} = \mathbf{r}$ are available.

For extension in the context of constitutive behavior, anisotropy must also be considered. As shown in[5], the isotropy assumption from the Naumenkonian homogeneity postulate [11] can loose validity. Therein, the orientation dependence of the three dimensional continuum [13, 15] is transferred to a planar continuum in the sense of Rychlewski [6] also. Furthermore, the extension of the harmonic decomposition [10] of the constitutive tensor to other symmetry groups proves to be practical.

The extensions listed here are summarized in Fig. 8.1. Here, the aspects considered in the context of present treatise are shown in light blue while framed in black.

Another aspect is the consideration of production-related residual stresses in the composite [17]. Within the framework of the theory developed here, the consideration of initial stresses in the form of the initial stress-resultants can be introduced in the constitutive relations. Furthermore, a thermodynamic generalization of the purely mechanical theory presented here should be considered. So far, thermal distortions as well as the influence of temperature fields have been disregarded. If, however, one considers the unsteady temperature effects acting in outdoor weathering, such an expansion is necessary in the context of the consideration of time-variant stresses of the individual components.

The theory of planar surface continua represents a special case with respect to the use of surface structures. The most general form represents a shell. Since

Anti-Sandwiches can also be produced as initially curved surfaces, an extension to curved surface continua is of interest. A first step in this direction was taken by Naumenko and Eremeyev [12], where the layered approach was extended to shallow shells. As an extension, the influence of thickness distortions and cross-sectional buckling may also be included. There are 6 and 7 parameter models available in the hierarchy of models for the numerical treatment of surface structures that could be applied for every layer independently. Finally, it should not be unmentioned that the theory of surface continua presented here is still controversial.

A validation of the theory presented here and the generated results by means of experimental investigations remains open. Although in [14] it was possible to make a comparison to the closed-form solution of a plate strip by means of a static load test, experimental comparisons to the eigenbehavior have not yet been made.

The underlying concept of the work has weaknesses. The theory of surface continua is very well suited to map the geometry of the deformation in the static load case as well as the frequencies of the eigenbehavior. These two dimensional theories basically have only an approximate character, since the underlying body is always physically three dimensional [9]. Due to the high degree of abstraction, the characteristic values for a strength analysis must be classified as untrustworthy, especially considering shear-soft structures. Here, stress and strain recovery will not result in measures known from Cauchy's theory. The coupling to the local structural analysis seems to be more purposeful and elegant [4] in this context. However, this coupling is associated with a considerable numerical effort. Ultimately, this leaves open the question of whether an a priori structural analysis based on a Cauchy continuum is not more effective and efficient, especially with regard to physical nonlinearities.

References

1. Altenbach H (1987) The direct approach in the theory of viscoelastic shells (in Russian). Habilitation thesis, Leningrad Polytechnic Institute
2. Altenbach H, Eremeyev V (2017) Thin-walled structural elements: classification, classical and advanced theories, new applications, pp 1–62. https://doi.org/10.1007/978-3-319-42277-0_1
3. Aßmus M, Köhl M (2012) Experimental investigation of the mechanical behavior of photovoltaic modules at defined inflow conditions. J Photonics Energy 2(1):1–11. https://doi.org/10.1117/1.JPE.2.022002
4. Aßmus M, Naumenko K, Altenbach H (2016) A multiscale projection approach for the coupled global-local structural analysis of photovoltaic modules. Compos Struct 158:340–358. https://doi.org/10.1016/j.compstruct.2016.09.036
5. Aßmus M, Nordmann J, Naumenko K, Altenbach H (2017) A homogeneous substitute material for the core layer of photovoltaic composite structures. Compos Part B: Eng 112:353–372. https://doi.org/10.1016/j.compositesb.2016.12.042
6. Blinowski A, Ostrowska-Maciejewska J, Rychlewski J (1996) Two-dimensional Hooke's tensors – isotropic decomposition, effective symmetry criteria. Arch Mech 48(2):325–345. http://am.ippt.pan.pl/am/article/view/v48p325/463
7. Bourbaki N (1971) Elemente der Mathematikgeschichte. Vandenhoek & Ruprecht, Göttingen
8. Föppl A (1907) Vorlesungen über technische Mechanik. B.G. Teubner, Leipzig

9. Koiter W (1969) Theory of thin shells. Springer, Heidelberg, chap Foundations and basic equations of shell theory: a survey of recent progress, pp 93–105. IUTAM Symposium Copenhagen 1967. http://www.springer.com/gp/book/9783642884788
10. Kowalczyk-Gajewska K, Ostrowska-Maciejewska J (2009) Review on spectral decomposition of Hooke's tensor for all symmetry groups of linear elastic material. Eng Trans 57(3–4):145–183. www.ippt.pan.pl/Repository/o451.pdf
11. Naumenko K, Eremeyev VA (2014) A layer-wise theory for laminated glass and photovoltaic panels. Compos Struct 112:283–291. https://doi.org/10.1016/j.compstruct.2014.02.009
12. Naumenko K, Eremeyev VA (2017) A layer-wise theory of shallow shells with thin soft core for laminated glass and photovoltaic applications. Compos Struct 178:434–446. https://doi.org/10.1016/j.compstruct.2017.07.007
13. Nye JF (1957) Physical properties of crystals: their representation by tensors and matrices. Oxford University Press, Ely House
14. Schulze SH, Pander M, Naumenko K, Altenbach H (2012) Analysis of laminated glass beams for photovoltaic applications. Int J Solids Struct 49(15):2027–2036. https://doi.org/10.1016/j.ijsolstr.2012.03.028
15. Voigt W (1966) Lehrbuch der Kristallphysik (mit Ausschluss der Kristalloptik). Springer, Wiesbaden. https://doi.org/10.1007/978-3-663-15884-4. Reproduktion des 1928 erschienenen Nachdrucks der ersten Auflage von 1910
16. von Kármán T (1910) Festigkeitsprobleme im Maschinenbau. Encyklopädie der mathematischen Wissenschaften IV:311–384
17. Zhang QZ, Shu BF, Chen MB, Liang QB, Fan C, Feng ZQ, Verlinden PJ (2015) Numerical investigation on residual stress in photovoltaic laminates after lamination. J Mech Sci Technol 29(2):655–662. https://doi.org/10.1007/s12206-015-0125-y

Appendix A
Rudiments of Tensor Calculus

This section is devoted to the fundamentals of the Tensor calculus. Through its mathematical structure, it forms the framework for the rational presentation in present treatise. Nevertheless, only the most necessary tools are to be explained at this point. For an exhaustive discussion of this topic, please refer to the literature, such as [2, 4, 6–8] or [1], among others.

The mechanical considerations refer to the Euclidean space. This space is selected as it has a metric, so provides measures for lengths and angles. This space is given a canonical orientation by a positively oriented base $\{\boldsymbol{e}_i\}$. This linear independent system of vectors is called orthonormal if the following applies.

$$\boldsymbol{e}_i \cdot \boldsymbol{e}_j = \begin{cases} 1 & \text{if} \quad i = j \\ 0 & \text{if} \quad i \neq j \end{cases} \tag{A.1}$$

First, second, and fourth-order tensors are introduced, giving priority to symbolic notation. In the following, these tensors are represented in component representation with respect to an orthonormal basis $\{\boldsymbol{e}_i\}$. The Einstein sum convention [3] applies here. Duplicate (dummy) indices are summed up.

$$\boldsymbol{a} = \sum_{i=1}^{3} a_i \boldsymbol{e}_i = a_i \boldsymbol{e}_i \tag{A.2}$$

$$\boldsymbol{A} = \sum_{i=1}^{3} \sum_{j=1}^{3} A_{ij} \boldsymbol{e}_i \otimes \boldsymbol{e}_j = A_{ij} \boldsymbol{e}_i \otimes \boldsymbol{e}_j \tag{A.3}$$

$$\boldsymbol{\mathcal{A}} = \sum_{i=1}^{3} \sum_{j=1}^{3} \sum_{k=1}^{3} \sum_{l=1}^{3} A_{ijkl}\, \boldsymbol{e}_i \otimes \boldsymbol{e}_j \otimes \boldsymbol{e}_k \otimes \boldsymbol{e}_l = A_{ijkl}\, \boldsymbol{e}_i \otimes \boldsymbol{e}_j \otimes \boldsymbol{e}_k \otimes \boldsymbol{e}_l \tag{A.4}$$

Usually the indices $i, j, k, l \in \{1, 2, 3\}$ are used in present explanations. The use of $\alpha, \beta, \gamma, \delta \in \{1, 2\}$ is analogous. Already introduced indirectly in Eq. (A.1), the Kronecker symbol (also Kronecker delta) is defined as follows.

M. Aßmus, *Structural Mechanics of Anti-Sandwiches*,
SpringerBriefs in Continuum Mechanics,
https://doi.org/10.1007/978-3-030-04354-4

$$\delta_{ij} = \begin{cases} 1 & \text{if} \quad i = j \\ 0 & \text{if} \quad i \neq j \end{cases} \tag{A.5}$$

It proves to be useful in the presentation of second- $\mathbf{1} = \delta_{ij}\boldsymbol{e}_i \otimes \boldsymbol{e}_j = \boldsymbol{e}_i \otimes \boldsymbol{e}_i$ or fourth-order metric tensors $\boldsymbol{\mathcal{I}} = \delta_{il}\delta_{jk}\boldsymbol{e}_i \otimes \boldsymbol{e}_j \otimes \boldsymbol{e}_k \otimes \boldsymbol{e}_l = \boldsymbol{e}_i \otimes \boldsymbol{e}_j \otimes \boldsymbol{e}_j \otimes \boldsymbol{e}_i$. In addition, the Levi-Civita or permutation symbol is also required. When considering three dimensions, it is triple indexed.

$$\epsilon_{ijk} = \begin{cases} +1, & \text{if } (i, j, k) \text{ is an even permutation of } (1, 2, 3) \\ -1, & \text{if } (i, j, k) \text{ is an odd permutation of } (1, 2, 3) \\ 0, & \text{if } (i, j, k) \text{ is no permutation of } (1, 2, 3) \end{cases} \tag{A.6}$$

The scalar product between first-, first- and second- and second-order tensors is defined as follows.

$$\boldsymbol{a} \cdot \boldsymbol{b} = a_i \, b_i \tag{A.7}$$

$$\boldsymbol{A} \cdot \boldsymbol{b} = A_{ij} b_j \, \boldsymbol{e}_i \tag{A.8}$$

$$\boldsymbol{A} \cdot \boldsymbol{B} = A_{ij} B_{kl} \boldsymbol{e}_i \otimes \boldsymbol{e}_i \cdot \boldsymbol{e}_k \otimes \boldsymbol{e}_l = A_{ij} B_{jl} \boldsymbol{e}_i \otimes \boldsymbol{e}_l \tag{A.9}$$

The scalar product between first-order tensors is commutative, but not between first- and second-order tensors, and that between fourth-order tensors. With the scalar product, e.g. the components of a tensor are determined.

$$A_{ij} = \boldsymbol{e}_i \cdot \boldsymbol{A} \cdot \boldsymbol{e}_j \tag{A.10}$$

Two vectors $\boldsymbol{a} \neq \mathbf{o}$ and $\boldsymbol{b} \neq \mathbf{o}$ are mutually orthogonal if their scalar product yields zero ($\boldsymbol{a} \cdot \boldsymbol{b} = 0$).

The double-scalar product between second-order tensors is defined as follows.

$$\begin{aligned} \boldsymbol{A} : \boldsymbol{B} &= A_{ij} \, B_{kl} \, \boldsymbol{e}_i \otimes \boldsymbol{e}_j : \boldsymbol{e}_k \otimes \boldsymbol{e}_l \\ &= A_{ij} \, B_{ji} \end{aligned} \tag{A.11}$$

Between a tensor fourth and a second-order tensors it can be determined as follows.

$$\begin{aligned} \boldsymbol{\mathcal{A}} : \boldsymbol{B} &= A_{ijkl} \, B_{mn} \, \boldsymbol{e}_i \otimes \boldsymbol{e}_j \otimes \boldsymbol{e}_k \otimes \boldsymbol{e}_l : \boldsymbol{e}_m \otimes \boldsymbol{e}_n \\ &= A_{ijkl} \, B_{lk} \boldsymbol{e}_i \otimes \boldsymbol{e}_j = \boldsymbol{F} \end{aligned} \tag{A.12}$$

The operation between two fourth-order tensors is determined as follows.

$$\begin{aligned} \boldsymbol{\mathcal{A}} : \boldsymbol{\mathcal{B}} &= A_{ijkl} \, B_{mnop} \, \boldsymbol{e}_i \otimes \boldsymbol{e}_j \otimes \boldsymbol{e}_k \otimes \boldsymbol{e}_l : \boldsymbol{e}_m \otimes \boldsymbol{e}_n \otimes \boldsymbol{e}_o \otimes \boldsymbol{e}_p \\ &= A_{ijkl} \, B_{lkop} \boldsymbol{e}_i \otimes \boldsymbol{e}_j \otimes \boldsymbol{e}_o \otimes \boldsymbol{e}_p = \boldsymbol{\mathcal{G}} \end{aligned} \tag{A.13}$$

The dyadic product between first-order tensors is as follows.

$$\boldsymbol{a}\otimes\boldsymbol{b} = a_i b_j \boldsymbol{e}_i \otimes \boldsymbol{e}_j = A_{ij}\boldsymbol{e}_i \otimes \boldsymbol{e}_j \qquad A_{ij} \neq A_{ji} \tag{A.14}$$

$$\boldsymbol{b}\otimes\boldsymbol{a} = b_j a_i \boldsymbol{e}_j \otimes \boldsymbol{e}_i = A_{ji}\boldsymbol{e}_i \otimes \boldsymbol{e}_j \qquad a_i b_j \neq b_j a_i \tag{A.15}$$

It is not commutative ($\boldsymbol{a}\otimes\boldsymbol{b} \neq \boldsymbol{b}\otimes\boldsymbol{a}$). The dyadic product between second-order tensors can be determined as follows.

$$\boldsymbol{A}\otimes\boldsymbol{B} = A_{ij}B_{kl}\boldsymbol{e}_i \otimes \boldsymbol{e}_j \otimes \boldsymbol{e}_k \otimes \boldsymbol{e}_l = C_{ijkl}\boldsymbol{e}_i \otimes \boldsymbol{e}_j \otimes \boldsymbol{e}_k \otimes \boldsymbol{e}_l \tag{A.16}$$

$$\boldsymbol{B}\otimes\boldsymbol{A} = B_{kl}A_{ij}\boldsymbol{e}_k \otimes \boldsymbol{e}_l \otimes \boldsymbol{e}_i \otimes \boldsymbol{e}_j = D_{ijkl}\boldsymbol{e}_i \otimes \boldsymbol{e}_j \otimes \boldsymbol{e}_k \otimes \boldsymbol{e}_l \tag{A.17}$$

Also, this link is not commutative. ($\boldsymbol{A}\otimes\boldsymbol{B} \neq \boldsymbol{B}\otimes\boldsymbol{A}$).

The transposition of a second-order tensor $\boldsymbol{A}^\top$ is defined by

$$\boldsymbol{a}\cdot\boldsymbol{A}^\top\cdot\boldsymbol{b} = \boldsymbol{b}\cdot\boldsymbol{A}\cdot\boldsymbol{a} \qquad \boldsymbol{A}^\top = A_{ji}\boldsymbol{e}_i \otimes \boldsymbol{e}_j \tag{A.18}$$

and the transposition of a tensor of fourth-order $\boldsymbol{\mathcal{A}}^\top$ by

$$\boldsymbol{A} : \boldsymbol{\mathcal{A}}^\top : \boldsymbol{B} = \boldsymbol{B} : \boldsymbol{\mathcal{A}} : \boldsymbol{A} \qquad \boldsymbol{\mathcal{A}}^\top = A_{klij}\boldsymbol{e}_i \otimes \boldsymbol{e}_j \otimes \boldsymbol{e}_k \otimes \boldsymbol{e}_l \,. \tag{A.19}$$

The cross product between two first-order tensors is defined as follows.

$$\begin{aligned}\boldsymbol{c} = \boldsymbol{a}\times\boldsymbol{b} &= a_i b_j \boldsymbol{e}_i \times \boldsymbol{e}_j \\ &= a_i b_j \epsilon_{ijk}\boldsymbol{e}_k \\ &= \|\boldsymbol{a}\|\,\|\boldsymbol{b}\| \sin\varphi\, \boldsymbol{e}_c \qquad \|\boldsymbol{a}\| = \sqrt{\boldsymbol{a}\cdot\boldsymbol{a}}\end{aligned} \tag{A.20}$$

The result is a vector $\boldsymbol{c}$ which is orthogonal to the plane spanned by $\boldsymbol{a}$ and $\boldsymbol{b}$. φ is the smaller angle between $\boldsymbol{a}$ and $\boldsymbol{b}$. The cross product is anti-commutative.

$$\boldsymbol{a}\times\boldsymbol{b} = -\boldsymbol{b}\times\boldsymbol{a} \tag{A.21}$$

Also of interest is the difference between polar and axial vectors. Due to the special consideration of rotations, axial vectors are used in the present work. Physically, this describes a vector whose length corresponds to the length of a circular line segment and whose direction describes the direction of rotation. The polar vector, on the other hand, is characterized by defining the magnitude and direction of a straight line segment [11].

The cross product between a second-order and a first-order tensor is defined by the following expression.

$$\boldsymbol{A}\times\boldsymbol{c} = (\boldsymbol{a}\otimes\boldsymbol{b})\times\boldsymbol{c} = \boldsymbol{a}\otimes(\boldsymbol{b}\times\boldsymbol{c}) = \boldsymbol{a}\otimes\boldsymbol{d} \tag{A.22}$$

This cross product has the following property.

$$\boldsymbol{A} \times \boldsymbol{c} = -\left[\boldsymbol{c} \times \boldsymbol{A}^\top\right]^\top \tag{A.23}$$

The vectorial invariant of a second-order tensor can be determined by the scalar cross product with the unit tensor $\mathbf{1} = \boldsymbol{e}_i \otimes \boldsymbol{e}_i$. This is often designated with an subscript cross as $\boldsymbol{A}_\times$, cf. [9, 10].

$$\begin{aligned}\boldsymbol{A}_\times &= \mathbf{1} \cdot \times \boldsymbol{A} \\ &= (\boldsymbol{e}_i \otimes \boldsymbol{e}_i) \cdot \times (A_{kl}\, \boldsymbol{e}_k \otimes \boldsymbol{e}_l) \\ &= A_{kl}\, \delta_{ik}\, \boldsymbol{e}_i \times \boldsymbol{e}_l \\ &= A_{ij}\, \boldsymbol{e}_i \times \boldsymbol{e}_j\end{aligned} \tag{A.24}$$

For a compact notation, the box product is introduced, which corresponds to the scalar cross product and has the following characteristics.

$$\mathbf{1} \cdot \times \boldsymbol{A} = \mathbf{1} \boxtimes \boldsymbol{A} = \boldsymbol{A} \boxtimes \mathbf{1} = -\mathbf{1} \boxtimes \boldsymbol{A}^\top = -\boldsymbol{A}^\top \boxtimes \mathbf{1} \tag{A.25}$$

If $\boldsymbol{A} \in \mathbb{Sym}$ holds true, the vectorial invariant is vanishing.

Every can tensor can be assigned its symmetric $\boldsymbol{A}^{\text{sym}}$ ($\boldsymbol{A} = \boldsymbol{A}^T$ resp. $\boldsymbol{b} \cdot \boldsymbol{A} = \boldsymbol{A} \cdot \boldsymbol{b}$) and anti(sym)metric (or skew) part $\boldsymbol{A}^{\text{skw}}$ ($\boldsymbol{A} = -\boldsymbol{A}^\top$ resp. $\boldsymbol{b} \cdot \boldsymbol{A} = -\boldsymbol{A} \cdot \boldsymbol{b}$).

$$\boldsymbol{A} = \boldsymbol{A}^{\text{sym}} + \boldsymbol{A}^{\text{skw}} \qquad \begin{aligned}\boldsymbol{A}^{\text{sym}} &= \tfrac{1}{2}\left[\boldsymbol{A} + \boldsymbol{A}^\top\right] \\ \boldsymbol{A}^{\text{skw}} &= \tfrac{1}{2}\left[\boldsymbol{A} - \boldsymbol{A}^\top\right]\end{aligned} \tag{A.26}$$

Skew tensors can also be represented as follows.

$$\boldsymbol{A}^{\text{skw}} = \boldsymbol{a} \times \mathbf{1} = \mathbf{1} \times \boldsymbol{a} \tag{A.27}$$

For fourth-order tensors, the symmetric portion of the unit tensor is of interest. For this purpose, the transposer $\mathcal{T}$ should be introduced first [2].

$$\mathcal{T} = \boldsymbol{e}_i \otimes \boldsymbol{e}_j \otimes \boldsymbol{e}_i \otimes \boldsymbol{e}_j \tag{A.28}$$

It maps all second-order tensors into their transpose. For application, the tensor $\boldsymbol{A} \in \mathbb{Lin}$ is used here.

$$\begin{aligned}\mathcal{T} : \boldsymbol{A} &= \boldsymbol{e}_i \otimes \boldsymbol{e}_j \otimes \boldsymbol{e}_i \otimes \boldsymbol{e}_j : A_{kl} \boldsymbol{e}_k \otimes \boldsymbol{e}_l \\ &= \delta_{jk} \delta_{il} A_{kl} \boldsymbol{e}_i \otimes \boldsymbol{e}_j \\ &= A_{ji} \boldsymbol{e}_i \otimes \boldsymbol{e}_j \\ &= \boldsymbol{A}^\top\end{aligned}$$

Thus, the symmetric part of the fourth-order unit tensor can be determined.

$$\mathcal{I}^{\text{sym}} = \frac{1}{2}\left[\mathcal{I} + \mathcal{T}\right] = \frac{1}{2}\left[\boldsymbol{e}_i \otimes \boldsymbol{e}_j \otimes \boldsymbol{e}_j \otimes \boldsymbol{e}_i + \boldsymbol{e}_i \otimes \boldsymbol{e}_j \otimes \boldsymbol{e}_i \otimes \boldsymbol{e}_j\right] \tag{A.29}$$

This tensor, also called symmetrizer, maps all second-order tensors $\boldsymbol{A} \in \mathbb{L}\text{in}$ into their symmetric part.

$$\begin{aligned}
\mathcal{I}^{\text{sym}} : \boldsymbol{A} &= \frac{1}{2}\left[\boldsymbol{e}_i \otimes \boldsymbol{e}_j \otimes \boldsymbol{e}_j \otimes \boldsymbol{e}_i + \boldsymbol{e}_i \otimes \boldsymbol{e}_j \otimes \boldsymbol{e}_i \otimes \boldsymbol{e}_j\right] : A_{kl} \boldsymbol{e}_k \otimes \boldsymbol{e}_l \\
&= \frac{1}{2}\left[\delta_{ik}\delta_{jl} A_{kl} \boldsymbol{e}_i \otimes \boldsymbol{e}_j + \delta_{jk}\delta_{il} A_{kl} \boldsymbol{e}_i \otimes \boldsymbol{e}_j\right] \\
&= \frac{1}{2}\left[A_{ij} \boldsymbol{e}_i \otimes \boldsymbol{e}_j + A_{ji} \boldsymbol{e}_i \otimes \boldsymbol{e}_j\right] \\
&= \frac{1}{2}\left[\boldsymbol{A} + \boldsymbol{A}^\top\right] \\
&= \boldsymbol{A}^{\text{sym}}
\end{aligned}$$

The symmetrizer of the planar surface continuum $\mathcal{P}^{\text{sym}}$ can be derived analogously by replacing the indices from $ijkl$ to $\alpha\beta\gamma\delta$.

Furthermore, each tensor can be decomposed additively into its dilatoric $\boldsymbol{A}^{\text{dil}}$ and deviatoric part $\boldsymbol{A}^{\text{dev}}$. For three dimensional problems, this is done as follows.

$$\boldsymbol{A} = \boldsymbol{A}^{\text{dil}} + \boldsymbol{A}^{\text{dev}} \qquad \begin{aligned} \boldsymbol{A}^{\text{dil}} &= \tfrac{1}{3}\left[\boldsymbol{A} : \boldsymbol{1}\right]\boldsymbol{1} \\ \boldsymbol{A}^{\text{dev}} &= \boldsymbol{A} - \boldsymbol{A}^{\text{dil}} \end{aligned} \tag{A.30}$$

For two dimensional problems, however, the following relationships apply by using the unit tensor of the surface continuum $\boldsymbol{P} = \boldsymbol{e}_\alpha \otimes \boldsymbol{e}_\alpha$.

$$\boldsymbol{B} = \boldsymbol{B}^{\text{dil}} + \boldsymbol{B}^{\text{dev}} \qquad \begin{aligned} \boldsymbol{B}^{\text{dil}} &= \tfrac{1}{2}\left[\boldsymbol{B} : \boldsymbol{P}\right]\boldsymbol{P} \\ \boldsymbol{B}^{\text{dev}} &= \boldsymbol{B} - \boldsymbol{B}^{\text{dil}} \end{aligned} \tag{A.31}$$

The inverse of a tensor is defined by the following relationships.

$$\boldsymbol{A}^{-1} \cdot \boldsymbol{A} = \boldsymbol{A} \cdot \boldsymbol{A}^{-1} = \boldsymbol{1} \qquad \left[\boldsymbol{A}^{-1}\right]^{-1} = \boldsymbol{A} \tag{A.32}$$

$$\mathcal{A}^{-1} : \mathcal{A} = \mathcal{A} : \mathcal{A}^{-1} = \mathcal{I}^{\text{sym}} \qquad \left[\mathcal{A}^{-1}\right]^{-1} = \mathcal{A} \tag{A.33}$$

The inverse of the second-order metric tensor coincides with the metric tensor itself ($\boldsymbol{1}^{-1} = \boldsymbol{1}$). For the fourth-order metric tensor this only applies to its symmetric part ($[\mathcal{I}^{\text{sym}}]^{-1} = \mathcal{I}^{\text{sym}}$). The determinant of a tensor is considered as criterion for its invertibility. The invertibility of any tensor $\boldsymbol{A}$ is guaranteed if $\det \boldsymbol{A} \neq 0$ holds. The determinants of second- and fourth-order tensors can be defined by their eigenvalues λ_i.

$$\det A = \prod_{i=1}^{N\!D} \lambda_i \qquad \det \mathcal{A} = \prod_{i=1}^{3N\!D} \lambda_i \tag{A.34}$$

The respective eigenvalues can be determined by the associated eigenproblem.

$$A \cdot c_i = \lambda_i c_i \qquad \mathcal{A} : C_i = \lambda_i C_i \tag{A.35}$$

Herein, c_i and C_i are first- and second-order eigentensors. The characteristic polynomials for the non-trivial solutions can be represented as follows.

$$\lambda^3 - I \; \lambda^2 + II \; \lambda - III = 0 \tag{A.36}$$

$$\lambda^9 - I\lambda^8 + II\lambda^7 - III\lambda^6 + IV\lambda^5 - V\lambda^4 + VI\lambda^3 - VII\lambda^2 + VIII\lambda - IX = 0 \tag{A.37}$$

Therein $I, II, III, \ldots, IX$ are principal invariants of A and $\mathcal{A}$, respectively. The above transition allows the solution with standard methods of linear algebra. Further hints for the solution of the eigenvalue problem of fourth-order tensors are presented in [5], while the solution of second-order tensors is described in detail in the textbooks mentioned at the outset of this annex. A tensor $A \in \mathbb{Lin}$ is called positive-definite if

$$b \cdot A \cdot b > 0 \qquad \forall\, b \neq \mathbf{o} \tag{A.38}$$

holds. Also, if $A \in \mathbb{Sym}$, then all eigenvalues are positive.

The Nabla operator is defined by the following relationship. With the aid of this, the divergence, gradient, and rotation of a tensor can be formed.

$$\nabla = e_i \frac{\partial}{\partial X_i} \qquad \begin{cases} \nabla \cdot & \text{(divergence)} \\ \nabla \otimes & \text{(gradient)} \\ \nabla \times & \text{(curl)} \end{cases} \tag{A.39}$$

In the present treatise the gradient is written as $\nabla\Box = \nabla \otimes \Box$ for the sake of brevity.

The scalar product of the Nabla operator with itself $\nabla \cdot \nabla$ gives the scalar-valued Laplace operator ∇^2. When applied to a twice differentiable scalar field, this operator assigns the divergence of its gradient to the scalar.

$$(\nabla \cdot \nabla) w = \nabla^2 w = \nabla \cdot (\nabla w) \tag{A.40}$$

The application in two dimensional Cartesian coordinates provides the following relation.

$$\nabla^2 w = \frac{\partial^2 w}{\partial X_1^2} + \frac{\partial^2 w}{\partial X_2^2} \tag{A.41}$$

For reasons of a consistent notation, the Laplace operator will not be denoted as Δ here, as commonly found in the literature.

For the conversion of surface into volume integrals and vice versa, this treatise uses the Gauss integral theorem (representative for the scientists Gauss, Ostrogradski, Stokes and Green) used. For explanation, V is a volume in $\mathbb{E}^3$ with its surface A and a surface normal $\boldsymbol{n}$. At this point, a generalized integral theorem is introduced to represent gradient, divergence and rotation theorems. The transformation of the integrals can thus be described as follows .

$$\int_V \nabla \circ \square \, \mathrm{d}V = \int_A \boldsymbol{n} \circ \square \, \mathrm{d}A \qquad \forall \, \square \equiv \{R \geq 1\} \quad \wedge \quad \circ = \begin{cases} \otimes \\ \cdot \\ \times \end{cases} \tag{A.42}$$

The symbol $\square$ stands here for a tensor field of order $R \geq 1$.

References

1. Altenbach J, Altenbach H (1994) Einführung in die Kontinuumsmechanik. Teubner, Stuttgart
2. Bertram A (2012) Elasticity and plasticity of large deformations: an introduction, 2nd edn. Springer, Berlin. http://dx.doi.org/10.1007/978-3-642-24615-9
3. Einstein A (1916) Die Grundlage der allgemeinen Relativitätstheorie. Annalen der Physik 354(7):769–822. http://dx.doi.org/10.1002/andp.19163540702
4. Haupt P (2002) Continuum mechanics and theory of materials, 2nd edn. Springer, Berlin. http://dx.doi.org/10.1007/978-3-662-04775-0
5. Itskov M (2000) On the theory of fourth-order tensors and their applications in computational mechanics. Comput Methods Appl Mech Eng 189(2):419–438. http://dx.doi.org/10.1016/S0045-7825(99)00472-7
6. Itskov M (2015) Tensor algebra and tensor analysis for engineers, 4th edn. Springer, Cham. http://dx.doi.org/10.1007/978-3-319-16342-0
7. Lai W, Rubin D, Krempl E (2010) Introduction to continuum mechanics, 4th edn. Butterworth-Heinemann, Oxford
8. Lebedev LP, Cloud MJ, Eremeyev VA (2010) Tensor analysis with applications in mechanics. World Scientific, New Jersey. http://dx.doi.org/10.1142/9789814313995
9. Legally M (1962) Vorlesungen über Vektorrechnung. Geest & Portig, Leipzig
10. Wilson EB (1901) Vector analysis, founded upon the lectures of G.W. Gibbs. Yale University Press, New Haven
11. Zhilin PA (2001) Vectors and second-rank tensors in three-dimensional space (in Russian). Nestor, St. Petersburg

Appendix B
Elastic Potential of Simple Materials

In the modeling of reversible, time-independent processes, the stress resultants are solely determined by the state of deformation [6]. When restricting to decoupled deformation states of a surface continuum, this leads to a unambiguous assignment.

$$\boldsymbol{N} = \boldsymbol{N}(\boldsymbol{G}) \qquad \boldsymbol{q} = \boldsymbol{q}(\boldsymbol{g}) \qquad \boldsymbol{L} = \boldsymbol{L}(\boldsymbol{K}) \tag{B.1}$$

The requirement of path independence is given when the kinetic measures can be derived by differentiating an elastic potential function W with respect to the conjugate kinematic quantities.

$$\boldsymbol{N} = \frac{\partial W}{\partial \boldsymbol{G}} \qquad \boldsymbol{q} = \frac{\partial W}{\partial \boldsymbol{g}} \qquad \boldsymbol{L} = \frac{\partial W}{\partial \boldsymbol{K}} \tag{B.2}$$

The strain energy function W depends only on the deformations [5]. It is a homogeneous quadratic function of $\boldsymbol{G}$, $\boldsymbol{g}$, and $\boldsymbol{K}$ [4].

$$W = \frac{1}{2}\left[\boldsymbol{G} : \boldsymbol{\mathcal{A}} : \boldsymbol{G} + \boldsymbol{K} : \boldsymbol{\mathcal{D}} : \boldsymbol{K} + \boldsymbol{g} \cdot \boldsymbol{Z} \cdot \boldsymbol{g}\right] \tag{B.3}$$

Therein, $\boldsymbol{\mathcal{A}}$ and $\boldsymbol{\mathcal{D}}$ are constitutive tensors of fourth-order, and $\boldsymbol{Z}$ is a constitutive tensor of second order. A discussion of the possible coupling of membrane strains, curvature changes, and more specifically with transverse shear strains, can be reviewed in e.g. [1, 2]. The strain energy should be positive definite.

$$W(\boldsymbol{G}, \boldsymbol{K}, \boldsymbol{g}) > 0 \qquad \forall \{\boldsymbol{G}, \boldsymbol{K}\} \neq \boldsymbol{0} \wedge \boldsymbol{g} \neq \mathbf{o} \tag{B.4}$$

Thus, the constitutive tensors must also be positive-definite. The following individual portions can be identified.

M. Aßmus, *Structural Mechanics of Anti-Sandwiches*,
SpringerBriefs in Continuum Mechanics,
https://doi.org/10.1007/978-3-030-04354-4

$$\begin{aligned} W_{\mathrm{M}} = W(\boldsymbol{G}) &= \frac{1}{2}\left[\boldsymbol{G} : \boldsymbol{\mathcal{A}} : \boldsymbol{G}\right] \\ &= \frac{Yh}{2(1+\nu)}\boldsymbol{G} : \boldsymbol{G} + \frac{Yh\nu}{2(1-\nu^2)}\left[\boldsymbol{G} : \boldsymbol{P}\right]^2 \\ &= G\,h\,\boldsymbol{G} : \boldsymbol{G} + \frac{1}{2}(B-G)h\left[\boldsymbol{G} : \boldsymbol{P}\right]^2 \end{aligned} \tag{B.5}$$

$$\begin{aligned} W_{\mathrm{B}} = W(\boldsymbol{K}) &= \frac{1}{2}\left[\boldsymbol{K} : \boldsymbol{\mathcal{D}} : \boldsymbol{K}\right] \\ &= \frac{Yh^3}{24(1+\nu)}\boldsymbol{K} : \boldsymbol{K} + \frac{Yh^3\nu}{24(1-\nu^2)}\left[\boldsymbol{K} : \boldsymbol{P}\right]^2 \\ &= G\,\frac{h^3}{12}\,\boldsymbol{K} : \boldsymbol{K} + \frac{1}{2}(B-G)\frac{h^3}{12}\left[\boldsymbol{K} : \boldsymbol{P}\right]^2 \end{aligned} \tag{B.6}$$

$$\begin{aligned} W_{\mathrm{S}} = W(\boldsymbol{g}) &= \frac{1}{2}\left[\boldsymbol{g} \cdot \boldsymbol{Z} \cdot \boldsymbol{g}\right] \\ &= \frac{\kappa Yh}{4(1+\nu)}\boldsymbol{g} \cdot \boldsymbol{g} \\ &= \frac{1}{2}\kappa\,G\,h\,\boldsymbol{g} \cdot \boldsymbol{g} \end{aligned} \tag{B.7}$$

For the forces and moments we obtain the following constitutive equations.

$$\boldsymbol{N} = \boldsymbol{\mathcal{A}} : \boldsymbol{G} = 2Gh\;\;\boldsymbol{G} + (B-G)h\;\;\left[\boldsymbol{G} : \boldsymbol{P}\right]\boldsymbol{P} \tag{B.8}$$

$$\boldsymbol{L} = \boldsymbol{\mathcal{D}} : \boldsymbol{K} = 2G\frac{h^3}{12}\boldsymbol{K} + (B-G)\frac{h^3}{12}\left[\boldsymbol{K} : \boldsymbol{P}\right]\boldsymbol{P} \tag{B.9}$$

$$\boldsymbol{q} = \boldsymbol{Z} \cdot \boldsymbol{g} \;\;= \kappa Gh\;\;\boldsymbol{g} \tag{B.10}$$

Equations (B.8)–(B.10) are valid only in the simplest case of material symmetries. Terms for the coupling of membrane and bending state as well as transverse shear state, as they are required in the generally anisotropic case as well as in the consideration of initially curved surfaces are not considered here. The stiffness tensors $\boldsymbol{\mathcal{A}}$, $\boldsymbol{\mathcal{D}}$, and $\boldsymbol{Z}$ can be given as follows.

$$\boldsymbol{\mathcal{A}} = 2Gh\;\;\boldsymbol{\mathcal{P}}^{\mathrm{sym}} + (B-G)\,h\;\;\boldsymbol{P}\otimes\boldsymbol{P} \tag{B.11}$$

$$\boldsymbol{\mathcal{D}} = 2G\frac{h^3}{12}\boldsymbol{\mathcal{P}}^{\mathrm{sym}} + (B-G)\,\frac{h^3}{12}\,\boldsymbol{P}\otimes\boldsymbol{P} \tag{B.12}$$

$$\boldsymbol{Z} = \kappa\,G\,h\boldsymbol{P} \tag{B.13}$$

Correlations to the engineering parameters Y and ν are as follows.

$$Y = \frac{4BG}{B+G} \qquad\qquad \nu = \frac{B-G}{B+G} \tag{B.14}$$

In the classical theories of thin-walled structural elements, representations of membrane stiffness (index M), bending stiffness (index B), and transverse shear stiffness (index S) have been manifested [8]. These are as follows.

$$D_{\mathrm{M}} = \frac{Y\,h}{1-\nu^2} = (B+G)\,h \quad D_{\mathrm{B}} = \frac{Y\,h^3}{12\left(1-\nu^2\right)} = (B+G)\,\frac{h^3}{12} \quad D_{\mathrm{S}} = \kappa\,G\,h \tag{B.15}$$

This leads to an alternative representation, as it is maintained e.g. in [7].

$$\boldsymbol{\mathcal{A}} = D_{\mathrm{M}}\,(1-\nu)\,\boldsymbol{\mathcal{P}}^{\mathrm{sym}} + D_{\mathrm{M}}\,\nu\,\boldsymbol{P}\otimes\boldsymbol{P} \tag{B.16}$$

$$\boldsymbol{\mathcal{D}} = D_{\mathrm{B}}\,(1-\nu)\,\boldsymbol{\mathcal{P}}^{\mathrm{sym}} + D_{\mathrm{B}}\,\nu\,\boldsymbol{P}\otimes\boldsymbol{P} \tag{B.17}$$

$$\boldsymbol{Z} = D_{\mathrm{S}}\,\boldsymbol{P} \tag{B.18}$$

It is trivial to show that Eqs. (B.11)–(B.13) or (B.16)–(B.18) can be transformed into Eq. (2.52).

Altenbach [3] introduces a further way of representation by grouping the bases while restricting to second-order tensors. This strategy follows differential-geometric thoughts. Due to the equivalence of these representations, however, this possibility should not be explicitly presented here. The representation in the Eqs. (B.16)–(B.18) competes with the representations in (2.52). While the projector representation is useful in terms of mathematical operations, the representation presented above characterized by engineering interpretations ($D_{\mathrm{M}}, D_{\mathrm{B}}, D_{\mathrm{S}}$) is advantageous in the context of the numerical treatment as presented in subsequent sections.

References

1. Altenbach H (1984) Analytische Modelle zur Beschreibung von in Dickenrichtung homogenen und inhomogenen dünnen Platten und Schalen. Zeitschrift für Angewandte Mathematik und Mechanik 64(10):M430–M431. http://dx.doi.org/10.1002/zamm.19840641003
2. Altenbach H (1985) Zur Theorie der inhomogenen Cosserat-Platten. Zeitschrift für Angewandte Mathematik und Mechanik 65(12):638–641. http://dx.doi.org/10.1002/zamm.19850651219
3. Altenbach H (1987) The direct approach in the theory of viscoelastic shells (in Russian). Habilitation thesis, Leningrad Polytechnic Institute
4. Altenbach H, Zhilin P (1988) A general theory of elastic simple shells (in Russian). Adv Mech (Uspekhi Mekhaniki) 11(4):107–148
5. Backhaus G (1983) Deformationsgesetze. Akademie Verlag, Berlin
6. Bertram A, Glüge R (2015) Solid mechanics: theory, modeling, and problems. Springer, Cham. http://dx.doi.org/10.1007/978-3-319-19566-7
7. Naumenko K, Eremeyev VA (2014) A layer-wise theory for laminated glass and photovoltaic panels. Compos Struct 112:283–291. http://dx.doi.org/10.1016/j.compstruct.2014.02.009
8. Timoshenko S, Woinowsky-Krieger S (1987) Theory of plates and shells, 2nd edn. McGraw-Hill, New York, (1st edn. 1959)

Appendix C
Vector-Matrix Formulation

To handle the boundary value problem numerically efficient, the basic equations must be transformed into a form readable by a computer algebra system. At first it seems beneficial to introduce a vector-matrix notation for the constitutive law. For this purpose, the schematic of the original Voigt notation [5] is used here. When applied to the three dimensional Cauchy continuum, the symmetry of stress and strain tensors as well as the main, left and right sub symmetry of the elasticity tensor are exploited. The reduction by a semi-circulating mnemonic rule, taking into account the orthonormal system, leads to a 6×1 vector representation for second-order tensors and a 6×6 matrix representation for fourth-order tensors. When restricting to orientation-independent material behavior, the following representation applies to linear constitutive relation $\boldsymbol{T} = \boldsymbol{\mathcal{C}} : \boldsymbol{E}$ known as Hookes law [2].

$$\mathbf{t} = \mathbf{C}\,\mathbf{e}$$
$$\begin{bmatrix} T_{11} \\ T_{22} \\ T_{33} \\ T_{23} \\ T_{13} \\ T_{12} \end{bmatrix} = \begin{bmatrix} C_{1111} & C_{1122} & C_{1122} & 0 & 0 & 0 \\ & C_{1111} & C_{1122} & 0 & 0 & 0 \\ & & C_{1111} & 0 & 0 & 0 \\ & & & C_{2323} & 0 & 0 \\ & & & & C_{2323} & 0 \\ \text{sym} & & & & & C_{2323} \end{bmatrix} \begin{bmatrix} E_{11} \\ E_{22} \\ E_{33} \\ 2E_{23} \\ 2E_{13} \\ 2E_{12} \end{bmatrix} \tag{C.1}$$

Herein, T_{ij} are stresses related to the Cauchy stress tensor, E_{kl} are linearized strains, and C_{ijkl} are material parameters of the constitutive tensor. This representation is mathematically inconsistent due to the following problems, cf. [4].

$$\mathbf{t}^{\top}\mathbf{t} \neq \boldsymbol{T} : \boldsymbol{T} \qquad \mathbf{e}^{\top}\mathbf{e} \neq \boldsymbol{E} : \boldsymbol{E} \qquad \mathbf{t}^{\top}\mathbf{e} = \boldsymbol{T} : \boldsymbol{E}$$

However, it has found wide application especially in dealing with numerical solution techniques. The factor 2 in the representation of the strain vector is used for the identical representation of the strain energy function in tensor and vector-matrix notation.

M. Aßmus, *Structural Mechanics of Anti-Sandwiches*,
SpringerBriefs in Continuum Mechanics,
https://doi.org/10.1007/978-3-030-04354-4

$$2W = \mathbf{t}^{\top}\mathbf{e} = \boldsymbol{T} : \boldsymbol{E}$$

By exploiting the symmetries of the membrane force tensor $\boldsymbol{N}$ and the moment tensor $\boldsymbol{L}$, the Voigt notation can also be applied to the structural mechanics problem presented in present work as introduced in [3]. The first-order transverse shear force tensor $\boldsymbol{q}$ is also converted into the vector notation. For the single layer, the kinetic measures can be represented as follows, using $\mathbf{s}$ as the global kinetic quantity for the sake of simplicity and indexed according to the loading case.

$$\mathbf{s}_{\mathrm{M}} = \left[N_{11}\ N_{22}\ N_{12}\right]^{\top} \tag{C.2}$$

$$\mathbf{s}_{\mathrm{S}} = \left[Q_{1}\ Q_{2}\right]^{\top} \tag{C.3}$$

$$\mathbf{s}_{\mathrm{B}} = \left[M_{11}\ M_{22}\ M_{12}\right]^{\top} \tag{C.4}$$

Analogous approach is used for the kinematic measures $\boldsymbol{G}$, $\boldsymbol{K}$, and $\boldsymbol{g}$ with $\mathbf{e}$ being used as a global kinematic variable for simplification.

$$\mathbf{e}_{\mathrm{M}} = \left[G_{11}\ G_{22}\ 2G_{12}\right]^{\top} \tag{C.5}$$

$$\mathbf{e}_{\mathrm{S}} = \left[g_{1}\ g_{2}\right]^{\top} \tag{C.6}$$

$$\mathbf{e}_{\mathrm{B}} = \left[K_{11}\ K_{22}\ 2K_{12}\right]^{\top} \tag{C.7}$$

Based on the constitutive tensors (B.16)–(B.18), a matrix notation can be introduced in an analogous manner as done in Hooke's law as shown in Eq. (C.1). Membrane, plate, and transversal shear stiffness are then represented as follows, with isotropy being assumed.

$$\mathbf{A} = \begin{bmatrix} A_{1111} & A_{1122} & 0 \\ & A_{2222} & 0 \\ \mathrm{sym} & & A_{1212} \end{bmatrix} \quad \mathbf{D} = \begin{bmatrix} D_{1111} & D_{1122} & 0 \\ & D_{2222} & 0 \\ \mathrm{sym} & & D_{1212} \end{bmatrix} \quad \mathbf{Z} = \begin{bmatrix} Z_{11} & 0 \\ \mathrm{sym} & Z_{22} \end{bmatrix} \tag{C.8}$$

Values of the coefficients $A_{\alpha\beta\gamma\delta}$, $D_{\alpha\beta\gamma\delta}$ and $Z_{\alpha\beta}$ can be derived directly from the Eqs. (2.53)–(2.55) in conjunction with Eq. (2.52). Based on this, the following representation is mostly to be found in the literature, cf. [1].

$$\begin{bmatrix} \mathbf{s}_{\mathrm{M}} \\ \mathbf{s}_{\mathrm{B}} \\ \mathbf{s}_{\mathrm{S}} \end{bmatrix} = \begin{bmatrix} \mathbf{A} & \mathbf{0} & \mathbf{0} \\ & \mathbf{D} & \mathbf{0} \\ \mathrm{sym} & & \mathbf{Z} \end{bmatrix} \begin{bmatrix} \mathbf{e}_{\mathrm{M}} \\ \mathbf{e}_{\mathrm{B}} \\ \mathbf{e}_{\mathrm{S}} \end{bmatrix} \tag{C.9}$$

Consequently, the matrices required in the FEM are to be given in terms of global variables (Indices $\circ$, Δ, c) for the three layered composite in vector-matrix form. The constitutive tensors of the global quantities can be introduced in matrix notation as follows, introducing $\mathbf{C}$ as a global stiffness quantity for the sake of simplicity.

$$\hat{\mathbf{C}}_{\mathrm{M}}^{K} = \begin{bmatrix} a_{\mathrm{M}}^{K} + 2b_{\mathrm{M}}^{K} & b_{\mathrm{M}}^{K} & 0 \\ a_{\mathrm{M}}^{K} & a_{\mathrm{M}}^{K} + 2b_{\mathrm{M}}^{K} & 0 \\ 0 & 0 & b_{\mathrm{M}}^{K} \end{bmatrix} \qquad \forall\, K \in \{\circ, \Delta, c\} \tag{C.10}$$

$$\hat{\mathbf{C}}_{\mathrm{B}}^{K} = \begin{bmatrix} a_{\mathrm{B}}^{K} + 2b_{\mathrm{B}}^{K} & b_{\mathrm{B}}^{K} & 0 \\ a_{\mathrm{B}}^{K} & a_{\mathrm{B}}^{K} + 2b_{\mathrm{B}}^{K} & 0 \\ 0 & 0 & b_{\mathrm{B}}^{K} \end{bmatrix} \qquad \forall\, K \in \{\circ, \Delta, c\} \tag{C.11}$$

$$\hat{\mathbf{C}}_{\mathrm{S}}^{K} = a_{\mathrm{S}}^{K} \begin{bmatrix} 1 & 0 \\ 0 & 1 \end{bmatrix} \qquad \forall\, K \in \{\circ, \Delta, c\} \tag{C.12}$$

Here, the following abbreviations have been introduced based on the engineering interpretations for membrane stiffness D_{M}, bending stiffness D_{B}, and transverse shear stiffness D_{S}.

$$a_{L}^{K} = \begin{cases} D_{L}^{t}\nu^{t} + D_{L}^{b}\nu^{b} & \text{if } K = \circ \\ D_{L}^{t}\nu^{t} - D_{L}^{b}\nu^{b} & \text{if } K = \Delta \\ D_{L}^{c}\nu^{c} & \text{if } K = c \end{cases} \qquad \forall L \in \{\mathrm{M}, \mathrm{B}\} \tag{C.13}$$

$$b_{L}^{K} = \begin{cases} \frac{1-\nu^{t}}{2} D_{L}^{t} + \frac{1-\nu^{t}}{2} D_{L}^{b} & \text{if } K = \circ \\ \frac{1-\nu^{t}}{2} D_{L}^{t} - \frac{1-\nu^{t}}{2} D_{L}^{b} & \text{if } K = \Delta \\ \frac{1-\nu^{t}}{2} D_{L}^{c} & \text{if } K = c \end{cases} \qquad \forall L \in \{\mathrm{M}, \mathrm{B}\} \tag{C.14}$$

$$a_{\mathrm{S}}^{K} = \begin{cases} D_{\mathrm{S}}^{t} + D_{\mathrm{S}}^{b} & \text{if } K = \circ \\ D_{\mathrm{S}}^{t} - D_{\mathrm{S}}^{b} & \text{if } K = \Delta \\ D_{\mathrm{S}}^{c} & \text{if } K = c \end{cases} \tag{C.15}$$

With the above representation, the generalized stiffness matrices can be specified.

$$\mathbf{C}_{\mathrm{MB}}^{\circ} = \begin{bmatrix} \hat{\mathbf{C}}_{\mathrm{M}}^{\circ} & \mathbf{0} & \mathbf{0} & \mathbf{0} \\ \mathbf{0} & \hat{\mathbf{C}}_{\mathrm{M}}^{\circ} & \mathbf{0} & \mathbf{0} \\ \mathbf{0} & \mathbf{0} & \hat{\mathbf{C}}_{\mathrm{B}}^{\circ} & \mathbf{0} \\ \mathbf{0} & \mathbf{0} & \mathbf{0} & \hat{\mathbf{C}}_{\mathrm{B}}^{\circ} \end{bmatrix} \tag{C.16}$$

$$\mathbf{C}_{\mathrm{MB}}^{\Delta} = \begin{bmatrix} \mathbf{0} & \hat{\mathbf{C}}_{\mathrm{M}}^{\Delta} & \mathbf{0} & \mathbf{0} \\ \mathbf{0} & \mathbf{0} & \mathbf{0} & \mathbf{0} \\ \mathbf{0} & \mathbf{0} & \mathbf{0} & \hat{\mathbf{C}}_{\mathrm{B}}^{\Delta} \\ \mathbf{0} & \mathbf{0} & \mathbf{0} & \mathbf{0} \end{bmatrix} \tag{C.17}$$

$$\mathbf{C}_{\mathrm{S}}^{\circ} = \begin{bmatrix} \hat{\mathbf{C}}_{\mathrm{S}}^{\circ} & \mathbf{0} \\ \mathbf{0} & \hat{\mathbf{C}}_{\mathrm{S}}^{\circ} \end{bmatrix} \tag{C.18}$$

$$\mathbf{C}_{\mathrm{S}}^{\Delta} = \begin{bmatrix} \mathbf{0} & \hat{\mathbf{C}}_{\mathrm{S}}^{\Delta} \\ \mathbf{0} & \mathbf{0} \end{bmatrix} \tag{C.19}$$

The zero matrices in the Eqs. (C.16) and (C.17) each possess three columns and rows, while the null matrices in Eqs. (C.18) and (C.19) have only two columns and rows each. The $\mathbf{B}$ matrices for combining the approximation of local continuous kinematic measures with the discrete degrees of freedom of the element are given as follows.

$$\mathbf{B}_{\mathrm{MB}} = \left[\mathbf{B}_{\mathrm{MB}_1}\ \mathbf{B}_{\mathrm{MB}_2}\ \ldots\ \mathbf{B}_{\mathrm{MB}_N}\right] \qquad \mathbf{B}_{\mathrm{MB}_i} = \left[\hat{\mathbf{B}}^{\circ}_{\mathrm{M}_i}\ \hat{\mathbf{B}}^{\Delta}_{\mathrm{M}_i}\ \hat{\mathbf{B}}^{\circ}_{\mathrm{B}_i}\ \hat{\mathbf{B}}^{\Delta}_{\mathrm{B}_i}\right]^{\top} \tag{C.20}$$

$$\mathbf{B}_{\mathrm{S}} = \left[\mathbf{B}_{\mathrm{S}_1}\ \mathbf{B}_{\mathrm{S}_2}\ \ldots\ \mathbf{B}_{\mathrm{S}_N}\right] \qquad \mathbf{B}_{\mathrm{S}_i} = \left[\hat{\mathbf{B}}^{\circ}_{\mathrm{S}_i}\ \hat{\mathbf{B}}^{\Delta}_{\mathrm{S}_i}\right]^{\top} \tag{C.21}$$

The sub measures introduced herein are given in the following matrices.

$$\hat{\mathbf{B}}^{\circ}_{\mathrm{M}_i} = \begin{bmatrix} N^i_{,1} & 0 & 0 & 0 & 0 & 0 & 0 & 0 & 0 \\ 0 & N^i_{,2} & 0 & 0 & 0 & 0 & 0 & 0 & 0 \\ N^i_{,2} & N^i_{,1} & 0 & 0 & 0 & 0 & 0 & 0 & 0 \end{bmatrix} \tag{C.22}$$

$$\hat{\mathbf{B}}^{\Delta}_{\mathrm{M}_i} = \begin{bmatrix} 0 & 0 & N^i_{,1} & 0 & 0 & 0 & 0 & 0 & 0 \\ 0 & 0 & 0 & N^i_{,2} & 0 & 0 & 0 & 0 & 0 \\ 0 & 0 & N^i_{,2} & N^i_{,1} & 0 & 0 & 0 & 0 & 0 \end{bmatrix} \tag{C.23}$$

$$\hat{\mathbf{B}}^{\circ}_{\mathrm{B}_i} = \begin{bmatrix} 0 & 0 & 0 & 0 & 0 & 0 & N^i_{,1} & 0 & 0 \\ 0 & 0 & 0 & 0 & 0 & -N^i_{,2} & 0 & 0 & 0 \\ 0 & 0 & 0 & 0 & 0 & -N^i_{,1} & N^i_{,2} & 0 & 0 \end{bmatrix} \tag{C.24}$$

$$\hat{\mathbf{B}}^{\Delta}_{\mathrm{B}_i} = \begin{bmatrix} 0 & 0 & 0 & 0 & 0 & 0 & 0 & 0 & N^i_{,1} \\ 0 & 0 & 0 & 0 & 0 & 0 & 0 & -N^i_{,2} & 0 \\ 0 & 0 & 0 & 0 & 0 & 0 & 0 & -N^i_{,1} & N^i_{,2} \end{bmatrix} \tag{C.25}$$

$$\hat{\mathbf{B}}^{\circ}_{\mathrm{S}_i} = \begin{bmatrix} 0 & 0 & 0 & 0 & N^i_{,1} & 0 & N^i & 0 & 0 \\ 0 & 0 & 0 & 0 & N^i_{,2} & -N^i & 0 & 0 & 0 \end{bmatrix} \tag{C.26}$$

$$\hat{\mathbf{B}}^{\Delta}_{\mathrm{S}_i} = \begin{bmatrix} 0 & 0 & 0 & 0 & 0 & 0 & 0 & 0 & N^i \\ 0 & 0 & 0 & 0 & 0 & 0 & 0 & -N^i & 0 \end{bmatrix} \tag{C.27}$$

The differential operators for membrane, bending, and transverse shear state as well as their auxiliary matrices are structured as follows.

$$\mathbf{D}_{\mathrm{MB}} = \left[\mathbf{D}^{\circ}_{\mathrm{M}}\ \mathbf{D}^{\Delta}_{\mathrm{M}}\ \mathbf{D}^{\circ}_{\mathrm{B}}\ \mathbf{D}^{\Delta}_{\mathrm{B}}\right]^{\top} \tag{C.28}$$

$$\mathbf{D}_{\mathrm{S}} = \left[\mathbf{D}^{\circ}_{\mathrm{S}}\ \mathbf{D}^{\Delta}_{\mathrm{S}}\right]^{\top} \tag{C.29}$$

The sub measures are structured as follows.

$$\mathbf{D}_{\mathrm{M}}^{\circ} = \begin{bmatrix} \frac{\partial}{\partial X_1} & 0 & 0 & 0 & 0 & 0 & 0 & 0 & 0 \\ 0 & \frac{\partial}{\partial X_2} & 0 & 0 & 0 & 0 & 0 & 0 & 0 \\ \frac{\partial}{\partial X_2} & \frac{\partial}{\partial X_1} & 0 & 0 & 0 & 0 & 0 & 0 & 0 \end{bmatrix} \tag{C.30}$$

$$\mathbf{D}_{\mathrm{M}}^{\Delta} = \begin{bmatrix} 0 & 0 & \frac{\partial}{\partial X_1} & 0 & 0 & 0 & 0 & 0 & 0 \\ 0 & 0 & 0 & \frac{\partial}{\partial X_2} & 0 & 0 & 0 & 0 & 0 \\ 0 & 0 & \frac{\partial}{\partial X_2} & \frac{\partial}{\partial X_1} & 0 & 0 & 0 & 0 & 0 \end{bmatrix} \tag{C.31}$$

$$\mathbf{D}_{\mathrm{B}}^{\circ} = \begin{bmatrix} 0 & 0 & 0 & 0 & 0 & \frac{\partial}{\partial X_1} & 0 & 0 & 0 \\ 0 & 0 & 0 & 0 & 0 & 0 & \frac{\partial}{\partial X_2} & 0 & 0 \\ 0 & 0 & 0 & 0 & 0 & \frac{\partial}{\partial X_2} & \frac{\partial}{\partial X_1} & 0 & 0 \end{bmatrix} \tag{C.32}$$

$$\mathbf{D}_{\mathrm{B}}^{\Delta} = \begin{bmatrix} 0 & 0 & 0 & 0 & 0 & 0 & 0 & \frac{\partial}{\partial X_1} & 0 \\ 0 & 0 & 0 & 0 & 0 & 0 & 0 & 0 & \frac{\partial}{\partial X_2} \\ 0 & 0 & 0 & 0 & 0 & 0 & 0 & \frac{\partial}{\partial X_2} & \frac{\partial}{\partial X_1} \end{bmatrix} \tag{C.33}$$

$$\mathbf{D}_{\mathrm{S}}^{\circ} = \begin{bmatrix} 0 & 0 & 0 & 0 & \frac{\partial}{\partial X_1} & 1 & 0 & 0 & 0 \\ 0 & 0 & 0 & 0 & \frac{\partial}{\partial X_2} & 0 & 1 & 0 & 0 \end{bmatrix} \tag{C.34}$$

$$\mathbf{D}_{\mathrm{S}}^{\Delta} = \begin{bmatrix} 0 & 0 & 0 & 0 & 0 & 0 & 0 & 1 & 0 \\ 0 & 0 & 0 & 0 & 0 & 0 & 0 & 0 & 1 \end{bmatrix} \tag{C.35}$$

The auxiliary matrices $\mathbf{A}_i \ \forall\, i \in \{1, \dots, 5\}$ for transforming the terms of virtual work into the vector-matrix notation are defined as follows.

$$\mathbf{A}_1 = \frac{1}{h^c} \begin{bmatrix} 0 & 0 & -2 & 0 & 0 & -(h^{\circ} + h^c) & 0 & -h^{\Delta} & 0 \\ 0 & 0 & 0 & -2 & 0 & 0 & -(h^{\circ} + h^c) & 0 & -h^{\Delta} \end{bmatrix} \tag{C.36}$$

$$\mathbf{A}_2 = \begin{bmatrix} 1 & 0 & 0 & 0 \\ 0 & 1 & 0 & 0 \end{bmatrix} \tag{C.37}$$

$$\mathbf{A}_3 = \left[\mathbf{I}\ \mathbf{0}\ \tfrac{1}{2}h^{\Delta}\mathbf{I}\ \tfrac{1}{2}h^{\circ}\mathbf{I}\right] \tag{C.38}$$

$$\mathbf{A}_4 = \frac{1}{h^c}\left[\mathbf{0}\ 2\mathbf{I}\ h^{\circ}\mathbf{I}\ h^{\Delta}\mathbf{I}\right] \tag{C.39}$$

$$\mathbf{A}_5 = \begin{bmatrix} 1 & 0 & 0 & 0 & 0 & 0 & 0 & 0 & 0 & 0 & 0 & 0 & 0 \\ 0 & 1 & 0 & 0 & 0 & 0 & 0 & 0 & 0 & 0 & 0 & 0 & 0 \\ 0 & 0 & 1 & 0 & 0 & 0 & 0 & 0 & 0 & 0 & 0 & -\frac{2}{h^c} & 0 \\ 0 & 0 & 0 & 1 & 0 & 0 & 0 & 0 & 0 & 0 & 0 & 0 & -\frac{2}{h^c} \\ 0 & 0 & 0 & 0 & 0 & 0 & 1 & 0 & 0 & 0 & 0 & 0 & 0 \\ \frac{1}{2}h^{\Delta} & 0 & \frac{1}{2}(h^{\circ} + h^c) & 0 & 0 & 0 & 0 & 0 & -1 & 0 & 0 & -\frac{h^{\circ}+h^c}{h^c} & 0 \\ 0 & \frac{1}{2}h^{\Delta} & 0 & \frac{1}{2}(h^{\circ} + h^c) & 0 & 0 & 0 & 1 & 0 & 0 & 0 & 0 & -\frac{h^{\circ}+h^c}{h^c} \\ 0 & 0 & 0 & 0 & \frac{1}{2}h^{\circ} & 0 & 0 & 0 & 0 & 0 & -1 & -\frac{h^{\Delta}}{h^c} & 0 \\ 0 & 0 & 0 & 0 & 0 & \frac{1}{2}h^{\circ} & 0 & 0 & 0 & 1 & 0 & 0 & -\frac{h^{\Delta}}{h^c} \end{bmatrix} \tag{C.40}$$

The unit $\mathbf{I}$ and the zero matrices $\mathbf{0}$ in Eqs. (C.38)–(C.39) each have three columns and rows. The auxiliary matrix for generating the mass matrix has nine columns and nine rows.

$$
\mathbf{H} = \begin{bmatrix} H_{11} & H_{12} & H_{13} & H_{14} & H_{15} & H_{16} & H_{17} & H_{18} & H_{19} \\ & H_{22} & H_{23} & H_{24} & H_{25} & H_{26} & H_{27} & H_{28} & H_{29} \\ & & H_{33} & H_{34} & H_{35} & H_{36} & H_{37} & H_{38} & H_{39} \\ & & & H_{44} & H_{45} & H_{46} & H_{47} & H_{48} & H_{49} \\ & & & & H_{55} & H_{56} & H_{57} & H_{58} & H_{59} \\ & & & & & H_{66} & H_{67} & H_{68} & H_{69} \\ & & & & & & H_{77} & H_{78} & H_{79} \\ & & & & & & & H_{88} & H_{89} \\ \text{sym} & & & & & & & & H_{99} \end{bmatrix} \tag{C.41}
$$

The parameters $H_{ij}\ \forall\, i, j \in \{1, \dots, 9\}$ in $\mathbf{H} \in \mathbb{Sym}$ are shorthand for the following components.

$$
\begin{aligned}
H_{11} &= H_{22} = H_{55} = \rho^{\circ} \\
H_{33} &= \rho^{\circ} - \frac{2}{3}\rho^{c} \\
H_{66} &= H_{77} = \left(\rho^{\circ} - \rho^{c}\right)\beta^{\circ} + \rho^{\vartriangle}\beta_{\vartriangle} + \frac{1}{12}\rho^{c}\left[(h^{\circ})^2 + 3\left(h^{\vartriangle}\right)^2\right] \\
H_{88} &= H_{99} = \left(\rho^{\circ} - \rho^{c}\right)\beta^{\circ} + \rho^{\vartriangle}\beta_{\vartriangle} + \frac{1}{12}\rho^{c}\left[\left(h^{\vartriangle}\right)^2 + 6\,(h^{\circ})^2\right] \\
H_{16} &= H_{27} = \left(\rho^{\circ} - \rho^{c}\right)\alpha^{\circ} + \rho^{\vartriangle}\alpha^{\vartriangle} + \frac{1}{2}\rho^{c}h^{\vartriangle} \\
H_{18} &= H_{29} = \left(\rho^{\circ} - \rho^{c}\right)\alpha^{\vartriangle} + \rho^{\vartriangle}\alpha^{\circ} + \frac{1}{2}\rho^{c}h^{\circ} \\
H_{36} &= H_{47} = \left(\rho^{\circ} - \rho^{c}\right)\alpha^{\vartriangle} + \rho^{\vartriangle}\alpha^{\circ} + \frac{1}{6}\rho^{c}h^{\circ} \\
H_{38} &= H_{49} = \left(\rho^{\circ} - \rho^{c}\right)\alpha^{\circ} + \rho^{\vartriangle}\alpha^{\vartriangle} + \frac{1}{6}\rho^{c}h^{\vartriangle} \\
H_{68} &= H_{79} = \left(\rho^{\circ} - \rho^{c}\right)\beta^{\circ} + \rho^{\vartriangle}\beta^{\circ} + \frac{1}{3}\rho^{c}h^{\circ}h^{\vartriangle}
\end{aligned}
$$

All other expressions in matrix (C.41) are identical to zero.

References

1. Altenbach H, Altenbach J, Kissing W (2004) Mechanics of composite structural elements. Springer, Berlin. http://dx.doi.org/10.1007/978-3-662-08589-9
2. Altenbach J, Altenbach H (1994) Einführung in die Kontinuumsmechanik. Teubner, Stuttgart
3. Aßmus M, Bergmann S, Eisenträger J, Naumenko K, Altenbach H (2017) Consideration of non-uniform and non-orthogonal mechanical loads for structural analysis of photovoltaic composite structures. In: Altenbach H, Goldstein RV, Murashkin E (eds) Mechanics for

materials and technologies, advanced structured materials, vol 46, Springer, Singapore, pp 73–122. http://dx.doi.org/10.1007/978-3-319-56050-2_4
4. Nordmann J, Aßmus M, Altenbach H (2018) Visualising elastic anisotropy: theoretical background and computational implementation. Contin Mech Thermodyn 30(4):689–708. http://dx.doi.org/10.1007/s00161-018-0635-9
5. Voigt W (1966) Lehrbuch der Kristallphysik (mit Ausschluss der Kristalloptik). Springer, Wiesbaden. http://dx.doi.org/10.1007/978-3-663-15884-4, Reproduktion des 1928 erschienenen Nachdrucks der ersten Auflage von 1910

Appendix D
Numerical Integration

The numerical integration is explicitly presented here, since artificial stiffening effects can be a problem of the finite element solution with the presented element. Such stiffening effects are characterized by the fact that the size sought with the numerical solution of the problem is smaller than that of the closed-form solution. The solution of structure mechanical problems is therewith characterized that the resulting values of the degrees of freedom are too small due to the too rigid mapping of the structure. Although convergence to the exact solution occurs with increasing mesh refinement, it does so much slower than with locking-free elements [7]. A distinction must be made between geometric and material locking effects. Geometric locking effects include plane shear locking, transverse shear locking, membrane locking, and trapezoidal locking, or curvature-thickness locking [4]. There is also a material locking effect, volumetric locking (also known as Poisson locking), which occurs predominantly in materials with a Poisson ratio near 0.5.

As is known from the literature, the element used in the present context tends to transverse shear locking [3, 6]. This stiffening effect becomes particularly relevant when bending states are studied in slender structures [1]. In this case, the shear stiffness is parasitic. To counter such problems, there are alternatives to full integration. Doherty et al. [2] introduced a procedure with a reduced order of integration (too low for an exact integration). In doing so, they limited themselves to defined stiffness terms, which from today's perspective is understood as selective integration. Reduced integration thus means the sub-integration of all stiffness terms. Selective integration is thus limited to the sub-integration of the terms associated to the transverse shear stiffness.

M. Aßmus, *Structural Mechanics of Anti-Sandwiches*,
SpringerBriefs in Continuum Mechanics,
https://doi.org/10.1007/978-3-030-04354-4

Table D.1 Gauss points and weights of Gauss–Legendre-quadrature [5]

	ξ_1^i	ξ_2^j	α^i, α^j
Full	$\xi_1^1 = -\sqrt{\frac{3}{5}}$	$\xi_2^1 = -\sqrt{\frac{3}{5}}$	$\alpha^1 = \frac{5}{9}$
	$\xi_1^2 = 0$	$\xi_2^2 = 0$	$\alpha^2 = \frac{8}{9}$
	$\xi_1^3 = +\sqrt{\frac{3}{5}}$	$\xi_2^3 = +\sqrt{\frac{3}{5}}$	$\alpha^3 = \frac{5}{9}$
Reduced	$\xi_1^1 = -\frac{1}{\sqrt{3}}$	$\xi_2^1 = -\frac{1}{\sqrt{3}}$	$\alpha^1 = 1$
	$\xi_1^2 = +\frac{1}{\sqrt{3}}$	$\xi_2^2 = +\frac{1}{\sqrt{3}}$	$\alpha^2 = 1$

In order to determine stiffness matrices, mass matrices and load vectors, it is necessary to integrate over the element surface. In the context of this work the Gauss-Legendre quadrature is used [7]. The analytic integration I of a function $f(\boldsymbol{\xi})$ over the two dimensional element $\mathrm{d}\Omega$ is defined as follows.

$$I = \int_{\Omega^e} f(\boldsymbol{\xi})\, \mathrm{d}\Omega^e = \int_{-1}^{1}\int_{-1}^{1} f(\boldsymbol{\xi})\, |\mathbf{J}(\boldsymbol{\xi})|\, \mathrm{d}\xi_1\, \mathrm{d}\xi_2 \tag{D.1}$$

For this problem, the integral can be defined as a weighted summation of the function values [5].

$$I \approx \sum_{i=1}^{NG_1}\sum_{j=1}^{NG_2} \alpha_1^i \alpha_2^j f(\xi_1^i, \xi_2^j) \tag{D.2}$$

The integration is performed in the interval $\xi_i \in [-1, 1]$. The function to be integrated is evaluated at the Gauss points ξ_i^j and multiplied by the weighting factors α_i^j. Here NG_i is the number of Gauss points in the direction considered. The coordinates ξ_i^i of the Gauss points as well as their weighting factors α_i^j for the planar SERENDIPITY element with quadratic shape functions are summarized in Table D.1 for complete and reduced integration. In addition, the different integration modes are visualized in Fig. D.1. However, this distinction only takes place in the integration of the stiffness terms.

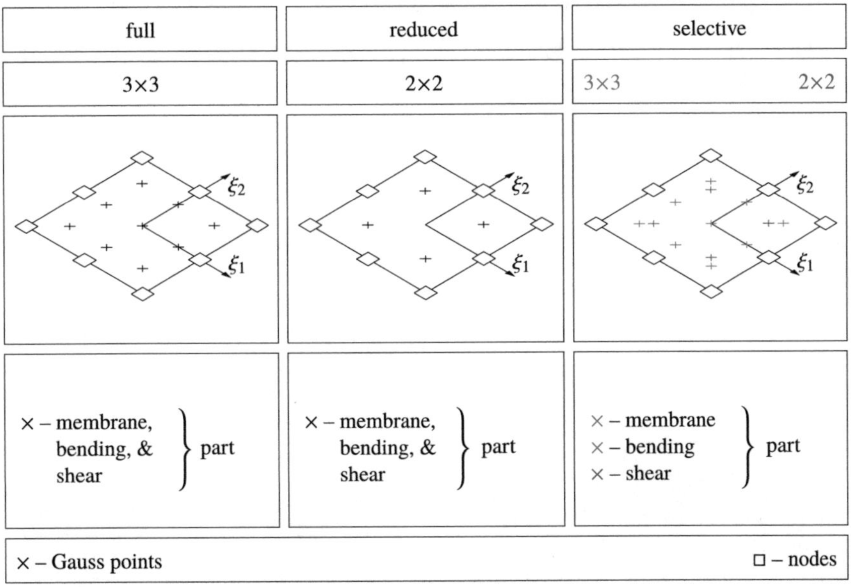

Fig. D.1 Usage of different Gauss points in the variation of integration types

References

1. Babuška I, Suri M (1992) On Locking and robustness in the finite element method. SIAM J Numer Anal 29(5):1261–1293. http://dx.doi.org/10.1137/0729075
2. Doherty WP, Wilson EL, Taylor RL (1969) Stress analysis of axisymmetric solids utilizing higher-order quadrilateral finite elements. Report sesm 69-3, Struct Eng Lab
3. Hughes TJR (1987) The finite element method. Linear static and dynamic finite element analysis. Prentice-Hall, Inc., Englewood Cliffs
4. Koschnick F (2004) Geometrische Locking-Effekte bei Finiten Elementen und ein allgemeines Konzept zu ihrer Vermeidung. Dissertation, Technische Universität München. http://nbn-resolving.de/urn:nbn:de:bvb:91-diss20041007006 24
5. Schwarz HR, Köckler N (2004) Numerische Mathematik, 5th edn. B.G. Teubner, Stuttgart. http://dx.doi.org/10.1007/978-3-322-96814-2
6. Szabó B, Babuška I (1991) Finite element analysis. Wiley, New York
7. Zienkiewicz OC, Taylor RL (2005) The finite element method for solid and structural mechanics, 6th edn. Elsevier Butterworth-Heinemann, Oxford

Printed in the United States
By Bookmasters